$56.00

D1074941

THE DANGERS OF LOW LEVEL RADIATION

The Dangers of Low Level Radiation

CHARLES SUTCLIFFE
Department of Economics
The University
Newcastle upon Tyne

Avebury
Aldershot · Brookfield USA · Hong Kong · Singapore · Sydney

© C.M.S. Sutcliffe 1987

All rights reserved. No part of this publication may be reproduced, stored in a retrieval system, or transmitted in any form or by any means, electronic, mechanical, photocopying, recording, or otherwise without the prior permission of Gower Publishing Company Limited.

Published by
Avebury
Gower Publishing Company Limited,
Gower House, Croft Road,
Aldershot
Hants GU11 3HR,
England

Gower Publishing Company,
Old Post Road, Brookfield,
Vermont 05036,
USA.

British Library Cataloguing in Publication Data

Sutcliffe, Charles
The dangers of low level radiation.
1. Ionizing radiation —— Physiological effect
I. Title
363.1'79 RA569

ISBN 0 566 05482 5

Printed in Great Britain by
Blackmore Press, Shaftesbury, Dorset

Contents

Preface

The impetus for my interest in the dangers of low level radiation goes back to the summer of 1981. As a newly elected member of Berkshire County Council, I was asked to present a petition to the Council signed by 5,500 people. This petition drew attention to the health risks from the discharge of radioactive liquids by the AWRE Aldermaston into the Thames at Pangbourne. I subsequently moved a motion, which was rejected, calling for an epidemiological study of cancer deaths in the area around Aldermaston.

It was early in 1982 that Brian Revell, then chairman of the Berkshire Anti Nuclear Campaign, suggested the production of a pamphlet on radiation risks within Berkshire. My projected contribution to this booklet just grew, and this book is the result. Whilst the initial motivation related to Berkshire, this book is world wide in its scope. Exposure to radiation occurs throughout the world and studies of the resulting health effects are available on a global basis.

I would like to thank the following people who have commented on various drafts of this book; Trevor Brown, Paul Cheshire, James Cutler, Tony Freke, David Gee, Dorian Owen, Professor Rotblat, Clif Sinclair, Jennie Snowdon and Tony Webb. I also wish to thank the secretaries who have produced various typescripts of this book and who have coped admirably; Mary Davies, Pat Elgar, Jill Turner, Meg Wells and Mary Wood.

Introduction

Exposure to radiation is a problem of increasing importance as radiation-based technologies proliferate. Everyone is exposed to some natural and man-made radiation. But, whilst exposure to natural radiation is largely unavoidable, exposure to man-made radiation is preventable. Society can choose the level of man-made radiation to which people are exposed, and a vital input to this decision is the health risks associated with exposure to low levels of radiation. Weighty reports by various national and international agencies are available which contain estimates of the health risks from exposure to small amounts of radiation (low level radiation). It will be argued that these official risk estimates are very imprecise, and it is possible that the dangers may be considerably greater than the official figures currently suggest. If the risks from exposure to radiation are higher than the official figures, then the exposure standards set using these risk estimates are required to be revised downwards. This is because the costs e.g. deaths from radiation exposure, have been revised upwards whilst the benefits are unchanged.

This book summarizes studies, published very largely in the last decade, which indicate that the risks from low level radiation are considerably higher than the official estimates. In consequence, only evidence that the risks are understated is presented. The empirical evidence in support of the existing risk estimates is amply covered in numerous official publications, whilst the studies which appear in this book are much less easily available. Throughout, an effort has been made to provide full references to the sources used, to permit the reader to consult the original empirical studies and follow-up the sources of information. Most of the studies showing that low level radiation risks are underestimated have either been subject to critical attack or ignored. Although some of the criticisms are unfair, considerable difficulties are involved in conducting epidemiological studies of the risks from low level radiation and, in consequence, some of the studies are indicative rather than conclusive. They should not be taken as unambiguously proving that low level radiation is much more dangerous than indicated by the official figures. Rather, they should be interpreted as showing that the official risk estimates are open to serious question.

Whilst exposure to radiation can cause a variety of adverse health effects, this book concentrates on cancer. Indeed, it further concentrates on cancer deaths rather than cancer incidence (fatal plus non-fatal cancers). This is because, so far, the studies have largely used cancer deaths. It does not mean that non-fatal cancers are unimportant, but merely that they have not been given their due prominence in the previous literature. The official reports have considered a large number of animal studies, but the studies considered in this book are almost entirely of humans. This is because animal experiments are thought to provide an unreliable guide to cancer risks in humans.

This book deals with the adverse health effects of low level radiation doses that have, and are, taking place. e.g. doses of under about five to ten rem per year (50-100 mGy), (see chapter 1 for an explanation of the terms rem and mGy). Such doses are caused by the normal functioning of the medical, industrial, military and power sectors. Possible hypothetical nuclear accidents are not directly considered, although there is discussion of the health effects of some nuclear accidents that have actually occurred. The health effects of possible accidents must be added to those already being caused by the routine functioning of nuclear establishments, medical exposure, nuclear weapons testing, natural radiation etc. The adverse health effects that would be caused by a nuclear attack are also not considered. This is because such an attack would involve large doses of radiation, and it is generally accepted that these are highly dangerous. The main controversy centres on the risks from exposure to small doses of radiation.

The first part of the book (chapters one to five) is mainly concerned with techniques for estimating the risks of human exposure to low level radiation, and the difficulties and uncertainties involved. The second part of the book (chapters six to ten) considers various categories of radiation exposure, and presents evidence of the dangers to humans. In particular, these chapters contain details of studies producing risk estimates above those which are officially accepted. Finally, chapter eleven considers a cost-benefit approach to radiation exposure.To help the reader, a glossary of terms is included at the end of this book.

1 Some basic radiobiology

1.1 Introduction

Radiobiology is the study of the effects of radiation on living things, and a full understanding of this subject (also called health physics) requires a knowledge of both nuclear physics and medicine. In addition, a knowledge of statistics is needed to conduct empirical studies of the biological effects of radiation. Researchers into radiobiology have tended to be experts in only one of these three areas, and have subsequently had to acquire some skills in the other two subjects.

This chapter attempts to introduce some of the relevant ideas from physics and medicine; a discussion of some of the statistical aspects being postponed until chapter 4. The nature of ionizing radiation, its measurement and the mechanisms by which it is thought to cause cancer are presented. This is followed by a discussion of the factors which influence the amount of biological damage done by radiation. Finally, various other matters such as latent periods and genetic effects are considered.

1.2 Ionizing Radiation

Radiation is generally divided into two categories: ionizing and non-ionizing radiation. Examples of non-ionizing radiation include light, lasers, radio-waves and micro-waves. Whilst non-ionizing radiation does not change the atomic structure of atoms, it can have adverse health effects e.g.sunburn. However, the health effects of non-ionizing radiation are not the subject of this book. Ionizing radiation is capable of changing the structure of atoms by removing electrons and leaving behind an atom with an electrical charge. Such a charged atom is called an ion. Ionizing radiation takes the form of either waves such as X-rays and gamma rays or particles such as alpha and beta particles and neutrons.

Ionizing radiation is emitted by radionuclides, which are simply elements that are in an unstable form i.e. radioactive. Some examples of radionuclides are uranium-238, caesium-137, iodine-131 and plutonium-239. The numbers after the names of these elements are the mass numbers and represent the number of protons plus the number of neutrons. The half-life of a radionuclide is the time taken for half the initial quantity (by weight) to decay and become some other substance. The products of this decay process are called the daughters of the initial radionuclide, and often they themselves are also radioactive. Since the daughters may also be radioactive, it is possible that the total amount of radioactivity emitted may actually rise over time, e.g. americium-242. The half-lives of some radionuclides are barium-140 13 days, caesium-137 30 years, carbon-14 5,730 years, iodine-131 8 days, plutonium-239 24,131 years, strontium-90 29 years and uranium-238 4,470 million years. An example of a decay chain is set out in table 1.1, where each element is the daughter of the preceding element, and the chain ends when the stable (i.e. non-radioactive) form of lead is reached. The forms of radiation most commonly emitted by radionuclides are alpha and beta particles and gamma rays. For example, one gramme of plutonium-239 emits about 2,000 million alpha particles per second, NRPB(1981). X-rays are usually produced by bombarding a metal target with electrons in a vacuum and were discovered in 1895 by W.K. Roentgen, who subsequently died of bone cancer in 1923. Henceforward the term 'radiation' will be used to denote ionizing radiation.

Radiation, whether in the form of particles or waves, is absorbed by collisions with the electrons of the matter through which it is passing. These collisions eject electrons from the atoms concerned, so ionizing the atom. The electrons which have been liberated are fast moving but are then slowed down by a series of ionizations and excitation processes. The ionizations involve collisions between the ejected electrons and the electrons in other atoms, leading to the creation of more ions and hence more fast moving electrons. An excitation imparts energy to an atom which is dissipated as heat without causing an ionization. Therefore, whether the radiation takes the form of particles or waves, in both cases the ionizations are ultimately caused by fast moving electrons, although the spatial distribution of the ionizations differs as between types of radiation. The reasons for differences in the spatial distribution of ionizations will be explained in the next section.

Table 1.1 Decay Chain for Radium-226

Sequence in Chain	Element	Half-life
1	Radium-226	1622 years
2	Radon-222	4 days
3	Polonium-218	3 months
4	Lead-214	27 months
5	Bismuth-214	20 months
6	Polonium-214	Under 1 second
7	Lead-210	22 years
8	Bismuth-210	5 days
9	Polonium-210	138 days
10	Lead-206	Stable

1.3 Linear Energy Transfer, Rads and Curies

Heavy particles, such as alpha particles, have a low (and variable) velocity and produce a zone of extremely high ionization and excitation in their wake which results in a rapid loss of energy. In consequence alpha particles do not penetrate more than a few millimeters into tissue. To use a simple analogy, alpha particles can be likened to a cannon ball which travels fairly slowly, has a high mass, does considerable damage to what it hits and does not travel far. X-rays can be likened to a high velocity rifle bullet which travels fast, has a low mass, makes only small holes in what it hits and travels a long way. Alpha particles cause very many ionizations in the tissue through which they have passed. In contrast X-rays, which are waves having no mass, travel at the speed of light and produce a very low density of ionization and excitation and so lose energy slowly. Hence they can easily penetrate through tissue and a path of low density ionizations will be created right through the tissue.

Differences in the rate of ionization and excitation between types of radiation have led to the concept of linear energy transfer (LET). This is defined as the energy deposition per path length of the original particle, and varies with the square of the charge and inversely with the velocity. Alpha particles with a double positive charge and a low velocity have a high LET, whilst X-rays and gamma rays, which have no charge and travel at the speed of light, have a low LET. It should be noted that since the speed of alpha and beta particles and neutrons can vary so can their rate of LET.

The quantity of radiation absorbed (whether in wave or particle form) was initially measured in roentgens, after the man who discovered X-rays in 1896. Subsequently, it was measured in rads, (radiation absorbed dose), where one roentgen equals 0.93 rad. The rad is the amount of energy absorbed per unit mass of matter. If human tissue is exposed to 400 rads, the amount of energy absorbed is sufficient to raise the temperature of the irradiated tissue by only 0.001 C. However, if such a dose were given to the entire body it would kill roughly half the people exposed, Gofman(1983,page 44). The rad has recently been replaced by the gray (named after a British scientist), where one gray equals 100 rads or one rad equals 0.01 Gy. However throughout this book rads not grays will be used. A table of conversion factors between the old and new units of measurement is set out in table 1.2. The rad only measures the radiation dose absorbed by some specified matter e.g. a human body and it is not a measure of the total quantity of radiation emitted by the radioactive source. This latter amount is measured in curies (named after Madame Curie the discoverer of radium, who died of leukaemia). The curie has recently been replaced by the becquerel (after the French scientist, Henri Becquerel, who discovered the radioactivity of uranium in November 1896), where one curie equals 37,000 million becquerels, but curies will continue to be used in this book.

Table 1.2 Conversion Between Old and New Units

	Old Unit	New Unit	Relationship
Activity	Curie	Becquerel	1 curie = 37,000 million becquerels
Absorbed dose	Rad	Gray (Gy)	100 rad = 1 gray 1 rad = 10 milligray (mGy)
Dose equivalent*	Rem	Sievert (Sv)	100 rem = 1 sievert 1 rem = 10 millisievert (mSv)

* See section 1.5 of this chapter for an explanation.

1.4 Biological Damage From Radiation

According to Morgan(1978) there are four things that can happen when radiation passes through tissue:-

(a) the radiation passes through or near cells without causing any damage.

(b) the radiation either kills a cell or renders it incapable of division i.e. sterilizes it.

(c) the radiation damages a cell but the damage is repaired.

(d) the radiation damages the nucleus of a cell and the cell survives to multiply in a deformed manner. This may then lead to the growth of a group of deformed cells which may be diagnosed as cancer.

There are two main ways in which radiation leads to biological damage. First, there is direct or physical action where the molecular bonds are broken by being hit by ejected electrons or by the original particle. Second, there is indirect or chemical action.

1.4.1 Physical Action

Target theory or hit theory has been developed which attempts to explain the adverse effects of radiation on the basis that ionization occurring in a discrete volume (the target) within the cell results in an adverse health effect. One, two or many 'hits' (ionizing events within the target) may be necessary to produce the adverse effect e.g. cancer, BEIR(1980, pp 523-4). If only a single hit is required then, for a given dose, low LET radiation will be more effective in producing adverse health effects than high LET radiation. This is because low LET radiation wastes little energy on 'hitting' the same cell more than once. If multiple hits are required, high LET radiation will be the more effective because its hits will be concentrated in a small area and so a multiple hit on the same target is much more likely.

The target is often taken to involve the strands of DNA. The nucleus of each human cell normally contains 23 pairs of chromosomes and each individual chromosome contains two strands of DNA(deoxyribonucleic acid). During cell division these strands of DNA separate so that each can be duplicated for transmission to the two daughter cells. In this way the daughter cells are exact copies of the original cell. If the strands of DNA are damaged or broken this can lead to the cell being unable to reproduce or to the new cells being different from their parent i.e. mutations. In order for a mutation to lead to the growth of a cancer the mutated cell must be such that it can survive and reproduce rapidly.

Pochin(1983, pp 93-4) suggests that if only one of the strands of DNA is broken it can be repaired within minutes. However correct repair is thought to be much less likely if both strands of DNA are broken at about the same position. Such a double break in the DNA strands might be caused by either a single hit or by two separate hits. But, as Pochin(1983, page 96) points out, a double break in the strands of DNA is not the only plausible hypothesis. An alternative theory is that if there are a large number of ionizations in a small area the local cell repair mechanisms become overwhelmed. In this case single breaks in the DNA may go unrepaired leading to cell mutations. The exact physical mechanism by which radiation causes cancer is still not clear. Indeed, Gofman(1983,page

239) has gone so far as to state that 'we do not know that DNA injury is at the bottom of cancer development at all'.

1.4.2 Chemical Action

Radiation can also have an indirect effect via the creation of free radicals which have an adverse chemical effect upon cells. A free radical is a grouping of atoms that normally exists in combination with other atoms, but can sometimes exist independently, and is generally highly chemically reactive. For example, dissolved oxygen in the cell fluid may capture a free electron to become the superoxide radical. This active form of oxygen can then lead to adverse chemical reactions in the cell such as damage to the DNA.

There are two mechanisms at work which cause an inverse relationship between the quantity of adverse chemical reactions and the density of the ionization. First, collisions between superoxide radicals convert them back into harmless oxygen. Thus the greater the density of superoxide radicals the greater is the chance of these recombination processes (collisions) occurring. Second, cell membranes produce an electric field that tends to attract negatively charged molecules such as the superoxide radicals. It has been found that when the ion concentrations are high the superoxide radicals are unable to reach the sensitive cell structures as easily as when the ion concentrations are low. So, again, dense ionization tends to reduce the chemical damage, Sternglass(1977).

The inverse relationship between the density of the ionization and the quantity of adverse chemical reactions has implications for the relative importance of the physical and chemical damage caused by high and low LET radiation. For radiation which produces dense ionization (e.g. alpha particles), the proportion of the adverse effects due to physical effects (as opposed to chemical effects) will be greater than for radiation which produces less dense ionization e.g. X-rays. There are also implications for the effects of varying the spatial distribution of the dose (irrespective of the type of radiation). If a dose of a fixed number of rads is given to a single area of tissue, the density of the ionization will be higher than if the same total dose had been distributed across more tissue. In consequence a concentrated dose will have less adverse chemical effects than a more widely distributed dose, Sternglass(1977).

1.5 Relative Biological Effectiveness and Rems

As outlined in section 3 of this chapter, types of radiation differ in their rates of LET. For example, alpha radiation produces many ionizations in a small area whilst the ionizations produced by X-rays are very much more dispersed. In section 4 some theories of the mechanisms by which radiation causes cancer were presented. These suggest that the rate of LET has an important effect upon cancer causation. Whilst LET can be measured, there is not a simple relationship between LET and can Hence there is no way of using physical measures of different kin radiation to produce a homogeneous measure of biological damag attempt has been made to allow for differences in the relative biologica effectiveness (RBE) of different types of radiation by using rems rather than rads to measure the radiation dose. The rem (or roentgen equivalent man) has recently been replaced by the sievert (named after R.M. Sievert, a Swedish radiologist) where one sievert equals 100 rems or one rem equals 0.01Sv.

The rem is a measure of the dose equivalent and may be defined as Rem = Rad x RBE where RBE is a conversion factor. The RBE factors have been standardized so that the RBE of therapeutic X-rays is equal to one i.e. one rad of X-rays equals one rem. The RBE values are usually taken as one for X-rays, gamma rays and beta particles, ten for fast neutrons and protons, and twenty for alpha particles. These RBE values imply that a dose of one rad of alpha particles is about twenty times as damaging as one rad of X-rays. Since rems measure the biological damage caused by different types of radiation, they are not amenable to precise physical measurement. However when considering the adverse health effects of radiation it makes sense to try to convert different types of radiation onto a common basis. These conversion factors (the RBE values) must themselves be estimated from the available data and so are subject to estimation error. For example, Charles et al (1983) have recently stated that 'it is clear that there is an urgent need to derive realistic values for the relative effectiveness of low doses of neutrons for cancer induction in humans'. This means that even though the dose in rads has been accurately measured, the dose in rems may be subject to considerable uncertainty.

The RBE can be expressed as Q x N, where Q is a physical (or quality) correction factor related to LET and N is a biological correction factor, Morgan(1978). Note that the LET of particles is not a constant since it varies inversely with their speed. The other correction factor, N, allows for additional factors which affect the amount of biological damage e.g. the age and sex of the subject, a non-uniform distribution of the dose, the dose rate, the type of animal etc. Differences in the sensitivity of various organs are handled by weighting factors and will be discussed in the next section. Since both Q and N will vary according to the circumstances, the RBE values (and therefore rems) are only approximations. There is also the additional complication of variations in the type of biological damage being considered e.g. cancer, chromosome aberrations etc. Whilst N is to allow for variations in the age and sex of the subject, the spatial distribution of the dose, the dose rate etc.; 'for the present the ICRP has assigned a value of unity to N', UNSCEAR(1982, page 46). This means that in calculating the RBE, the International Commission on Radiological Protection (ICRP) has ignored the effects of variations in these factors.

1.6 Effective Dose Equivalent

An important source of variation in the biological damage done by a given amount of radiation is the radiosensitivity of the tissue involved. For example, lung tissue will suffer considerably more damage than will the thyroid when each receives the same number of rads of the same type of radiation. To allow for differences in radiosensitivity, a set of weighting factors has been developed. If the whole body is subject to a uniform dose of radiation, the proportion of the total biological damage to the body that occurs in each organ can be estimated. For example, in 1981 the US Environmental Protection Agency (EPA) proposed the use of the following; lung 16%, breast 20%, red bone marrow 16%, bone surfaces 3%, thyroid 4%, skin 1% and the remainder 40%.

Often the dose of radiation is not the same to all organs of the body e.g. a dental X-ray will deliver a much smaller dose to the chest. If a single organ receives a dose of a given number of rems the relevant weighting factor can be used to convert this dose into an effective dose equivalent (also measured in rems). This represents the whole body dose which would produce the same biological damage as the actual dose to the organ or tissue concerned. For example, if the exposure is a dose of five rems (50 mSv) to the breast and three rems (30 mSv) to the lungs, with no other parts of the body receiving any extra radiation, the effective dose equivalent is 1.11 rems (11.1 mSv) i.e. $5 \times 0.15 + 3 \times 0.12$. The use of an effective dose equivalent (which is often abbreviated to dose) is a very convenient way of standardizing doses to different organs. However the weighting factors must be estimated from real world data and so are subject to dispute. The effective dose equivalent depends upon two sets of estimates which may or may not be accurate. First, there is the estimation of the RBE which is used to convert rads into rems, and second there is the estimation of the weighting factors which convert the dose in rems into a whole body equivalent. This assumes that the dose in rads has itself been accurately measured.

The effective dose equivalent provides a measure of the dose to an individual. The dose to a group of people is called the collective effective dose equivalent, and is just the sum of the effective dose equivalents for each member of the group. The collective effective dose equivalent , which may be abbreviated to collective dose, is measured in man-rems (or man-sieverts).

1.7 Some Factors Influencing RBE

A number of factors have been identified as influencing the relative biological effectiveness of radiation, in addition to LET. These are the age and sex of the subject, the dose rate, the type of animal involved and the spatial distribution of the dose.

(a) *Age*. The age of the subject can have an important influence upon the effectiveness of a given amount of radiation in inducing cancer. If the cells of the irradiated tissue are dividing very infrequently, damage to the DNA of some of these cells is much less likely to produce cancer than if the cells are dividing very rapidly. Therefore young children, whose cells are dividing more rapidly than those of adults, are at greater risk. In particular, it has been found that the foetus is the most vulnerable to radiation damage, and especial care must be taken to avoid X-raying pregnant women; see chapter 6 for a fuller discussion.

BEIR(1980, page 33) states that there is now considerable evidence that adults who are old when irradiated have a greater increase in their cancer rate than do adults who are younger when irradiated, or at least develop cancer sooner. Beentjes et al(1980) have pointed out that, whilst more recent data indicates that risk increases with age, in 1977 the ICRP stated that risk *decreased* with age. Beentjes et al attribute the increasing risk to a decline with age in the efficiency of the immune system. Of course, since exposure to radiation can take several decades before cancer is diagnosed, an elderly person may die from some other cause first and so for those over say 65 the risks will drop.

(b) *Sex*. There is evidence to suggest that women are more susceptible to radiation induced cancer than are men. BEIR(1980) present tables of their estimates of the risk of cancer following radiation exposure, with separate risk factors for men and women. Using an absolute risk model (see chapter 2) these show that for females the risk from exposure to low LET radiation is two to two and a half times that for males. This may in part be because women suffer from cancer of the cervix, breast and ovary; although men are subject to cancer of the prostate and testis. Whatever the reason, it appears that women are more vulnerable to radiation and this must be allowed for in any analysis.

(c) *Dose Rate*. The rate at which a given dose of radiation is delivered is thought to have an important effect on the amount of biological damage that is done. A fixed amount of rads can be delivered at a high dose rate or the same total number of rads can be given at a low dose rate. In some exposures e.g. the taking of a single X-ray, the entire dose will be delivered very quickly and so the dose rate is high. In other cases the dose rate may be low, with the total dose being spread out over months or years. Dose fractionation occurs when a given total dose is administered in relatively small doses daily or at longer intervals e.g. the taking of a series of medical X-rays. Dose protraction occurs when a given total dose is delivered continuously over a relatively long period e.g. cosmic radiation. Whilst it is widely accepted that variation in the dose rate is an important factor, opinions differ as to both the direction and magnitude which such variations in the dose rate can have upon the amount of biological damage.

A common argument is that a low dose rate permits the damage caused by radiation to be repaired. For, example, if one strand of DNA is damaged it can be repaired before subsequent radiation damages the matching DNA strand. If the radiation were delivered at a high dose rate both strands of DNA may be damaged simultaneously with subsequent repair being impossible. This view leads to the conclusion that, if anything, an increase in the dose rate should tend to increase the risk of cancer. Thus Pochin(1983, page 102) states that 'a given dose will be considerably less dangerous if delivered during a month than during a minute'. Pochin's argument relies on the fact that as the dose rate is increased so the density of the ionization is increased. This means that the simultaneous breaking of two DNA strands at the same point by physical action is more likely. However, Sternglass(1977) has argued that there is an inverse relationship between the quantity of indirect damage via chemical effects and the density of the ionization. Since, for a given type of radiation, there is a positive relationship between the dose rate and the density of the ionization, Sternglass argues that the amount of damage done by indirect effects increases as the dose rate drops. Thus the risk of physical effects causing cancer decreases as the dose rate drops, whilst the risk of indirect effects causing cancer increases when the dose rate drops. This leads to the question as to which effect is larger, and hence whether a reduction in the dose rate causes the risk of cancer to rise or fall.

A number of authors have made a distinction between high and low LET radiation when considering the effects of variations in the dose rate. For high LET radiation, such as alpha particles, there is some evidence suggesting that as the rate at which a given dose is delivered falls, the cancer risks rise. Mays(1973) reported that shortly after the Second World War about 2,000 German patients were given repeated injections of radium-224, an alpha emitter, as a treatment for tuberculosis and ankylosing spondylitis. It was found that as the time span of the injections was increased the incidence of bone sarcomas for a given total dose rose. The number of bone sarcomas per rad was roughly doubled by a five fold lengthening of the time period over which the injections of radium-224 were given. Hence high LET radiation was found to be more dangerous when delivered at a slow rate. This conflicts with the view that a lower dose rate permits repair mechanisms to operate. It implies that the risks of high LET radiation based upon high dose rates understate the risks at low dose rates e.g. alpha emitting particles lodged in the lungs; Mays et al (1978).

Petkau(1972) has considered the effects of altering the dose rate of low LET radiation. He studied the total dose of X-rays to which a phospholipid membrane surrounded by water had to be exposed before it broke. When the water contained a small quantity of radioactive sodium salt and the dose rate was 26 rads (260 mGy) per minute a total dose of 445 rads (4450 mGy) was required. He found that if the dose rate were reduced to only 0.1 rads (1 mGy) per minute the cumulative dose required to rupture the membrane dropped to only three rads (30 mGy). This is just 0.7% of the total dose required at the higher dose rate. Although these results are not based upon human data, they indicate that for low LET a lowering of the dose rate increases the biological damage.

This result conflicts with the conclusions of the National Council on Radiological Protection (NCRP), as reported by Charles et al (1983). The NCRP concluded that for low LET radiation a reduction in the dose rate can mean a reduction in the risk of cancer of between two and ten times. However this conclusion is largely based on animal studies, not on human data. If a reduction in the dose rate of low LET does lead to a reduction in the risk, an allowance must be made for this when using high dose rate data to estimate the risks of low LET exposure. The NCRP call such an allowance the dose rate effectiveness factor (DREF), and values of somewhere between two and five have been used by various standard setting bodies, Charles et al (1983). This means that risks estimated from high dose rate exposure are divided by the DREF to give an estimate of the risk at low dose rates. However, not only the magnitude but also the direction of the DREF values is uncertain, so that dividing risk estimates by numbers such as five is questionable. In the view of Charles et al (1983) for low LET radiation 'the NCRP review of dose rate effect factors illustrates the paucity of human data on which this can be based and a case can clearly be made for not using a DREF where a conservative and easily defensible risk factor is required'.

(d) *Type of Animal.* Whilst experimentation on humans has been ruled out, many scientific studies have been conducted using experimentation on animals and insects e.g. mice, rats, flies, beagle dogs etc. However, the results of such experiments may provide only a poor guide to the effects of radiation on humans. Dr. David Salsburg, a statistician employed by Pfizer Laboratories in Connecticut, has investigated the accuracy of animal experiments in determining whether various chemicals cause cancer. He found that, of nineteen known human carcinogens, only seven have been shown to cause cancer when fed to rodents. He concluded that instead of conducting experiments on animals to determine whether a chemical causes cancer in humans one might as well flip a coin, Heneson(1983). This suggests that the type of animal concerned can make a difference to the risk of radiation inducing cancer. Because of the uncertainties in translating the results of animal studies into risk estimates for humans, this book is concerned almost exclusively with human studies. Even for human studies care must be exercised in interpreting the results because of differences in those studied e.g. the healthy worker effect (see chapter 8), patients who already have some serious disease (see chapter 6) and Japanese who are still alive some years after 1945 (see chapter 7).

(e) *Spatial Distribution.* Rather than deliver an equal dose throughout some tissue or organ e.g. the lungs, the dose may be concentrated in just a small area. For example, if a particle of plutonium lodges in the lungs it will irradiate only the lung tissue within a few centimetres with alpha particles. This means that whilst the dose to the organ as a whole may be fairly low, some areas of tissue will be receiving a much higher dose and others a low dose. If the dose-response curve is linear (see chapter 2) the distribution of the dose does not matter, but otherwise the distribution of the dose will affect the cancer risk.

1.8 Cancer

Cancer is a malignant tumour that rapidly invades and replaces the surrounding tissue. Cancerous cells can also produce secondary growths some distance from the main tumour. As can be seen from table 1.3, which is taken from OPCS(1981), cancer can cause death by damaging a wide variety of parts of the body. The mechanisms by which cancer is caused are not fully understood. A wide variety of substances have been identified as carcinogens, with different carcinogens tending to cause different types of cancer. For example, cigarette smoking tends to cause lung cancer. As will be seen in later chapters, irradiation of a particular organ tends to cause cancer in that organ; so that uranium miners who breath in radioactive material tend to get lung cancer, whilst women who receive chest X-rays tend to get breast cancer. Some types of cancer are called 'radiosensitive cancers', and are those where a uniform dose of radiation to the whole body tends to produce the largest proportionate increase e.g. leukaemia.

Whilst roughly 20% of the UK population die from cancer, evidence to be considered later, suggests that exposure to radiation above background levels is responsible for only a small fraction of total cancer deaths. Both the studies of the health effects of radiation and the standard setting organizations have tended to focus on cancer deaths. Whilst many types of cancer have a high fatality rate some do not e.g. thyroid and skin cancer, and the human trauma of a non-fatal cancer means that consideration should be given to the causation of non-fatal cancers. Table 1.4, which is based on Gofman(1983, pp 220-221), shows that there is considerable variation in the mortality rate between cancers in different parts of the body and that, on average, 55% of cancers in the US are fatal.

Table 1.3 Number of Cancer Deaths in England and Wales in 1978

Type of Cancer	Males	Females	Total
Lung	26760	7588	34348
Breast	92	11915	12007
Stomach	6698	4799	11497
Large Intestine (Colon)	4324	6057	10381
Rectum	3202	2847	6049
Pancreas	3008	2692	5700
Prostate	4730	0	4730
Bladder	2939	1304	4243
Ovary	0	3765	3765
Oesophagus	2085	1623	3708
Leukaemia	1832	1540	3372
Brain	1305	874	2179
Cervix	0	2153	2153
Kidney	1083	648	1731
Other Cancers of the Uterus	0	1562	1562
Buccal Cavity and Pharynx	927	591	1518
Multiple Myeloma	758	732	1490
Gallbladder	426	691	1117
Lymphosacoma	532	430	962
Larynx	615	166	781
Malignant Melanoma of the Skin	334	429	763
Hodgkin's Disease	398	268	666
Liver	380	198	578
Cancer of the Skin	271	224	495
Bone	260	205	465
Connective Tissues	202	156	358
Testis	233	0	233
All Cancers	68107	58681	126788

The only completely successful treatment for cancer is the surgical removal of the entire cancer, and the earlier this is done the better. Chemotherapy i.e. drug treatment, is also used but the chemicals kill not only the cancerous cells but also normal cells. Radiotherapy can be used to kill cancerous cells in specific areas within the patient's body e.g. the use of high doses of X-rays to very specific parts of the body to kill deep-seated cancers.

Table 1.4 The Mortality/Incidence Ratio for Different Types of Cancer in the US in 1980

Type or Site of the Cancer	Percentage
Endocrine Glands	15
Eyes	22
Genital Organs	31
Buccal Cavity and Pharynx	35
Urinary Organs	35
Bone, Connective Tissue and Skin	47
Other Blood and Lymph Tissues	53
Digestive Organs	57
Leukaemia	71
Respiratory System	81
Brain and Central Nervous System	82
All Other Sites	90
Total	55

1.9 Internal and External Radiation

An important distinction is between external and internal radiation, since if radioactive material is inhaled its potential harmfulness is multiplied many times. External radiation occurs when the source of the radiation is located outside the body e.g. X-rays from a machine or cosmic rays from outer space. Such radiation typically irradiates all, or large parts, of the body and is usually low LET radiation like X-rays or gamma rays.

Doses of external radiation can be measured by ionization chambers, where the electrically charged ions produced within a volume of gas generate an electric current that indicates the size of the dose at each moment in time. Ionizing chambers are unable to record the LET of the radiation involved. A simpler way of measuring external radiation is film badges. Since radiation causes photographic plates to become clouded, people who are likely to be exposed to external radiation can wear a badge which contains a strip of film. After a week or so the film is developed and the degree of blackening indicates the size of the cumulative dose to which the badge has been exposed. If the badge is worn on the chest it will not necessarily measure the external dose to the hands, head or lower body of the wearer; which may be higher. Even when the dose is uniform to all parts of the body, there are problems in the use of badges to measure the dose and GMBATU(1983) reports that there can be 90% errors in the measurement of the dose. This type of badge can be modified to distinguish, in a crude way, between different types of radiation. This is done by designing the holder so that there are different thicknesses of plastic protecting different parts of the radiation sensitive material. High LET radiation will be unable to penetrate the thicker areas of plastic so the material underneath will be mainly affected by low LET radiation, Pochin(1983,pp 50-52).

Instead of using badges containing film, badges with a thin strip of thermoluminescent material can be used i.e. thermoluminescent dosimeter (TLD). Some of the atoms of such a material are displaced when it is exposed to radiation. The cumulative quantity of radiation is measured by heating the material and measuring the amount of light given off. An American study found that some TLD's were totally insensitive to the type of radiation for which they were purchased, Bertell(1985, page 54).

Internal radiation takes place when the source of the radiation is located within the body e.g. particles of plutonium dust in the lungs or thorium in the blood. Sources of internal radiation are particles of radioactive material that enter the body by being inhaled, swallowed, injected etc. The risks from internal radiation are influenced by the extent to which radioactive substances are absorbed into the body from the stomach, gut or lung; how long they remain within the body before being excreted; and in which organs they lodge. In chapter 10 there is a discussion of the estimates of the extent to which radioactive material is absorbed through the gut. The ease with which radioactive material is absorbed into the body depends upon its physical and chemical state. Different types of radioactive material tend to concentrate in different organs. For example, iodine concentrates in the thyroid gland and so radioactive iodine also builds up in the thyroid. Some types of radioactive material such as strontium-90 accumulate in the bones, whilst others like caesium-137 are deposited in muscle tissue.

The dose of internal radiation to which a person has been exposed is considerably more difficult to measure than the dose of external radiation. Where a person is occupationally exposed to contaminated air, air samplers can be used to estimate the resulting internal dose. However, there are dangers that the dose to a worker's lungs may be drastically underestimated. Static air samplers do not provide an accurate measure of the radioactive content of the air breathed by a worker, and personal air samplers are necessary. The ICRP has stated that for short term samples, static air samplers understate the dose by between 100 and 1000 times, ICRP(1969, page 11).

Once a person has received an internal dose it is more difficult to measure. The dose from LET radiation can be measured by surrounding the person's body with sensitive instruments that record the radiation being emitted from the subject's body i.e. a whole body count. This measurement, which takes place in a specially constructed room where all other sources of radiation are minimized, records the current rate of emission. Since high LET radiation is absorbed by a few centimetres of tissue, this technique can only measure radiation that passes through tissue and out of the body e.g. low LET radiation. The counters which are used to measure the radiation enable an estimate to be made as to which radioactive material is emitting the radiation. Another way of estimating the internal dose involves the measurement of the rate at which the person is excreting radioactive material. This is done by analyzing samples of urine. However, there is not a simple relationship between the urinary excretion rate and the current internal dose. Other methods of estimating the internal dose include analysis of breath, sweat, nasal secretions and faecal samples, Pochin(1983, pp 52-57). The measurement of internal doses is both more difficult and more unreliable than the measurement of external doses, and many epidemiological studies have used measurements of just the external dose.

Once radioactive material is inside the body it will continuously irradiate the surrounding body tissue until either it is excreted or it decays to some non-radioactive substance. Some radioactive material may never be excreted as it is incorporated into the person's bones etc. If such material has a half-life of several decades or more, the person will be subject to a continuous additional dose of radiation for the rest of his or her life. In contrast, exposure to most forms of man-made external radiation is for very short periods of time e.g. having an X-ray taken.

1.10 Latent Period

Some of the health effects from exposure to radiation occur shortly afterwards. Thus exposure to very high levels of radiation causes death within about a day. However, other health effects (e.g. cancer) occur very much later. For this reason such health effects are called 'late effects'. Since for any exposed individual radiation may or may not cause cancer, it is also termed a 'stochastic effect'. For leukaemia the latent period can vary from a few years to over thirty years. The average period between radiation exposure and the diagnosis of leukaemia is twelve years. For other forms of cancer the latent periods are longer, with the average lag between exposure and diagnosis being over twenty years, Pochin(1983, pp 126-127).

The length of the latent periods are difficult to estimate for a number of reasons:-

(a) they are subject to considerable variation from person to person,

(b) follow-up periods of over thirty years are required if a latent period of this length is to be detected. However, most data sets are for shorter periods, and so they cannot detect such long latent periods, and,

(c) for some solid cancers the latent period may be longer for persons exposed to radiation when they are younger, BEIR(1980, page 30).

1.11 Genetic Effects

An important distinction is between somatic and genetic health effects. Somatic effects are those health effects which are manifested in the exposed subjects e.g. cancer. Genetic effects are those which appear in the descendents of the exposed subjects e.g. the malformation of children conceived after their parent has been exposed to radiation. Radiation damage to a foetus will be defined as somatic damage, since the foetus itself is directly damaged by the radiation. Such damage can appear as cancer in the child (as discussed in chapter 6), or as a stillbirth (see chapter 9). It can also lead to teratogenic effects i.e. malformations and abnormalities in the child.

This book is almost entirely concerned with the somatic effects of radiation. This is not because genetic effects are unimportant, but because the epidemiological evidence on the risks of genetic damage is even thinner than it is for somatic risks. Because genetic risks only become apparent after one, two or more subsequent generations, a very long time period is needed to study genetic effects in human populations. In consequence, epidemiological evidence on genetic risks using human populations is very scarce and is likely to remain so for many years. This means that the estimates of the genetic risks to humans are based very largely on animal studies, Pochin(1983, page 120). However, as argued in chapter 4, studies using fruit flies, mice, spiderwort plants etc. may be a poor guide to the genetic risks in humans. Whilst genetic risks are not covered in this book their omission does not mean that they are unimportant. Currently the genetic risks are thought to be something like a quarter to three quarters of the somatic risks. Almost everywhere that cancer risks are discussed in this book there will also be a genetic risk which is additional to the cancer and other somatic risks.

Genetic damage is thought to occur in the same ways as somatic damage. The cell's DNA is damaged so that when it divides a mutation is produced. If the mutation occurs in a somatic cell the consequence may be cancer, whilst if the damage is to a germ cell the result may be deformities and abnormalities in subsequent generations.

1.12 Summary

When considering the health effects of ionizing radiation, it is important to distinguish between different types of radiation. Such differences can be allowed for by the use of RBE coefficients which depend upon the radiation's LET, the dose rate and various other features of the tissue irradiated e.g. sex and age of the subject, the organ irradiated, the type of animal and the spatial distribution of the dose. However, in calculating RBE coefficients, the ICRP used only the differences in LET. So the other factors which influence the amount of biological damage done by radiation are not allowed for in the RBE coefficients.

The precise mechanisms by which radiation causes cancer are not known. This means that theoretical models cannot be relied upon to calculate accurately the number of cancers which a dose of radiation will induce. Empirical estimates are needed, and these studies face substantial difficulties e.g. measuring the quantity and type of additional doses of internal and external radiation received. This book concentrates upon a particular late stochastic effect of radiation i.e. cancer. Another important late stochastic effect, genetic damage, is not explicitly considered because of the absence of human data on the risks. But almost everywhere that cancer risks are discussed there will also be a genetic risk.

2 Dose-response curves

2.1 Introduction

A prerequisite to the setting of radiation safety standards is the estimation of the relationship between the radiation dose and the consequent adverse health effects. Such a relationship is called the dose-response or dose-effect curve. Throughout this book the main adverse health effect to be considered is cancer. A very important step in setting radiation safety standards is the estimation of the dose-response curve, and there is still considerable controversy over its shape. In the first section of this chapter the assumption that the dose response curve is linear will be examined and some alternatives considered. The second section discusses the concept of a threshold dose, and it is argued that there is no such thing as a safe dose of radiation. Some reasons why the dose-response curve is concave are presented in section three, together with some factors suggesting that the true dose-response curve lies above the estimated dose-response curve. The third section considers the choice between using absolute and relative risk models. Finally, some quotations on the likely concavity of the dose-response curve and on the inadequacy of the available data are presented.

2.2 The Dose-Response Curve and the Linearity Assumption

In order to estimate the dose-response curve, data on the carcinogenic effects of different levels of radiation are necessary. If the dose-response curve is known to be linear then data for at least two radiation levels are necessary, whilst if the shape of the dose-response curve is unknown many more observations are needed for its accurate estimation.

There is a major difficulty in estimating the dose-response curve since (a) the shape of the curve cannot be constrained to be linear so that many observations are needed, and (b) what was thought to be authoritative data is available for only one part of the curve. These observations are taken from the effects of the atomic bombs dropped on Japan and, to a lesser extent, on those ankylosing spondylitis patients given high doses of radiation for therapeutic purposes. They provide high dose/high risk observations, and everyone is agreed that high doses of radiation are highly dangerous. This means that researchers have effectively one observation, and this is on the high dose portion of the dose-response curve. The validity of the Japanese data has recently been questioned and a re-evaluation is currently being undertaken, see chapter 5. So at the moment, even the high dose observation is open to doubt.

If the curve is linear at least one more observation is necessary for the specification of the dose-response curve. In order to obtain one additional observation at the low dose end of the curve it is commonly assumed that no additional dose of radiation implies no additional risk of cancer i.e. the dose-response curve passes through the origin (this assumption will be discussed in the next section). It is then officially assumed that the dose-response curve is a straight line relationship through these two points, i.e. line 1 in figure 2.1. However this linearity assumption, which is crucial to the way in which radiation safety standards have been set, has been disputed. It has been argued that in fact the true dose-response curve is concave i.e. line 2 in figure 2.1. There is a variety of empirical evidence for this point of view. For example, Gofman (1983) presents evidence for a concave dose-response curve for lung cancer in uranium miners, and for breast cancer, leukaemia and total cancer deaths amongst the survivors of the Hiroshima and Nagasaki bombs. Lindop (1978?) also draws attention to the concave dose-reponse curve for lung cancer at Hiroshima, whilst Smith et al (1982) found a concave dose-response curve for leukaemia amongst a group of patients given a single course of X-rays. Further evidence supporting a concave dose-response curve will be presented in subsequent chapters. If the dose-response curve is concave, the risk from a low dose of radiation is, in reality, B in figure 2.1 and not C as implied by the linearity assumption. In consequence the official safety standards may be much too lax for low doses.

The use of a linear dose-response curve is very convenient. Indeed Mole (1975a) suggests the linear hypothesis is 'almost, if not absolutely essential for radiological protection in practice'. It means that the cancer risks of a given total dose are independent of its distribution amongst individuals. Thus a dose of 1 rad (10 mGy) each to one million people carries the same risks as a dose of 1,000 rads (10,000 mGy) each to 1,000

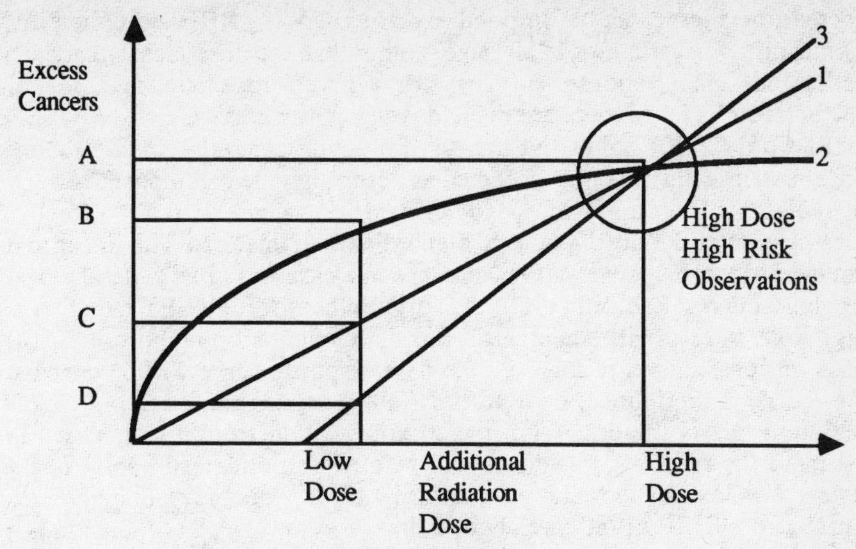

Figure 2.1 Dose-Response Relationships

people. The linear hypothesis also means that the risks of an additional dose are independent of the previous radiation doses received by that person. Hence the risks of exposing someone to an extra rad are the same whether they have previously received a cumulative dose of 100 rad (1,000 mGy) or 1 rad (10 mGy). In addition, the linear hypothesis means that the spatial distribution of the dose within an organ is of no importance. Whether the dose is concentrated in a few 'hot spots' or evenly distributed throughout the organ, the response is identical. Thus the linearity assumption massively simplifies the setting of radiation safety standards, and abandoning it would create very considerable problems for calculating the additional cancer risks to a population. Knowledge of the distribution of the dose between individuals, the previous cumulative dose of each individual and the spatial distribution within organs would now be required. So it is not surprising that those responsible for radiological protection cling to the very convenient linear hypothesis.

However, the linear hypothesis does not remove the necessity for any other assumptions. The chosen form of the dose-response curve is fitted to a set of real world data to estimate its parameters. But there may be some important differences between those individuals in the sample who have been exposed to high as opposed to low doses, e.g. dose-rate, age, sex, the organs irradiated, the type of radiation etc. If such differences in the data exist, their effects upon the response must be eliminated. This is a very difficult task since, as was stated in chapter 1, the correction factors may themselves vary e.g. with the size of the dose, dose rates and organs irradiated. Therefore some simple constant cannot be used as the correction factor, as is implied by the use of RBE co-efficients. Controlling for variations in all factors other than the dose means that each estimated dose-response curve is for a given dose rate, age, sex of subjects, etc. Hence there are many dose-response curves.

Radford (1980) points out that the following general equation for the dose-response curve is now accepted as describing the carcinogenic effects of radiation: $E = (aD + bD^2)\exp(-cD - dD^2)$, where E is the effect or response, D is the dose and a, b, c and d are parameters. The linear and concave dose-response curves are special cases of this general dose-response curve. Radford (1980) then argues that for high LET radiation at high dose rates d will equal zero. For the upward sloping portion of the dose response curve, if $b/a < c/2$ the dose response curve will be concave e.g. curve 2 in figure 2.1. Radford (1980) argues that for high LET radiation this is the case, and linear extrapolation from higher doses to estimate the risks at low doses will result in an underestimation of the risks. For low LET radiation Radford (1980) concludes that the risk estimates in BEIR (1980) are about 5 times too low, i.e. that radiation is 5 times more dangerous than claimed by BEIR.

If both the c and d parameters equal zero the dose-response curve becomes linear-quadratic, i.e. $E = aD + bD^2$ whilst if, in addition, b equals zero the dose response curve is quadratic, i.e. $E = bD^2$.

The type of dose-response curve which is fitted to the data can have a very important influence on the resulting risk estimates. Land (1980) fitted various mathematical specifications of the dose-resoponse curve to the Nagasaki bomb data and found that the estimates of the excess risk of leukaemia from exposure to one rad (10 mGy) varied by over 150 times as between a linear and a quadratic dose-resoponse curve. Thus the choice of model can have more influence on the estimate of excess risk at low doses than the data themselves, Land (1980).

2.3 The Dose-Response Curve and the Threshold Dose

In the past, when setting radiation safety standards, it has been argued that there is a radiation dose below which there are no carcinogenic effects. Thus instead of going through the origin, the dose-response curve intersects the horizontal axis at some positive radiation dose: the threshold dose. The effects of introducing a threshold dose for the case where the dose-response curve is linear are shown as line 3 in figure 2.1 This shows that for a low dose the estimated risk is only D, whilst a linear relationship with no threshold produces a higher risk of C, and a concave curve through the origin gives an even higher risk of B.

If there is a threshold dose and the radiation safety standard is set at or below this dose then exposure to smaller doses of radiation is harmless.

This means that provided the radiation dose received by any individual does not exceed the threshold level no cancers will be induced, i.e. a safe dose exists. An implication of this view is that a given population exposure to radiation should be spread amongst as many people as possible e.g. the practice of 'burning out' radiation workers described in chapter 8. Indeed, many official statements concerning the dangers of low level radiation can be best understood in terms of a 'threshold dose' mentality. For example, as late as 1970 the U.S. Atomic Energy Commission was claiming that doses of radiation within the permissible levels were harmless, Tamplin et al (1970), whilst in '1969 one "expert" seriously proposed a safe threshold, with respect to carcinogensis, at 1,000 rads' (10,000 mGy), quoted from Gofman (1983, page 244).

According to Morgan (1980), prior to about 1960 most experts accepted that there was a threshold dose for cancer induction by radiation. But since then the empirical studies have supported the view that there is **no** threshold, and between about 1960 and 1970 the threshold hypothesis was discarded by the Atomic Energy Commission (AEC), Biological Effects of Ionizing Radiation (BEIR), United Nations Scientific Committee on the Effects of Atomic Radiation (UNSCEAR), ICRP, EPA, NCRP, Federal Radiation Council and the Bureau of Radiological Health. Thus the concept of a threshold dose for radiation has now been generally rejected. In the report of the Windscale (Sellafield) Inquiry, (Parker (1978)) Justice Parker wrote that "it is common ground that all additional radiation is harmful to some extent" (page 8) and on page 44 he rejected the view that a threshold dose exists. This means that there is no such thing as a safe dose of radiation. **Any** radiation is dangerous, no matter how small the dose.

An example of the studies that have led to the rejection of a threshold dose is the study by Furcinitti et al (1979). They conducted an experiment on cultured human kidney cells to determine whether a threshold dose exists. They subjected each cell to a measured dose of gamma radiation in the range 20 to 90 rads (200 to 900 mGy). Each cell was then cultured to see if it had survived the radiation dose. For the three lowest radiation doses the experiment was repeated over 10,000 times to try to accurately determine the proportion of cells that survived. They then estimated a survival curve i.e. the relationship between the proportion of cells surviving irradiation and the dose of radiation they received. Two possible survival curves are illustrated in figure 2.2. Curve 1 has a threshold dose since 100% survive doses below some positive dose whilst curve 2 has no threshold as any positive dose reduces the survival rate to below 100%.

If there is a threshold dose the slope of the survival curve will be zero at zero dose, i.e. curve 1, whilst if there is no threshold the slope will be negative at zero dose, i.e. curve 2. The authors found clear statistical evidence that there is **no** threshold dose, i.e. curve 2 is correct, so that any

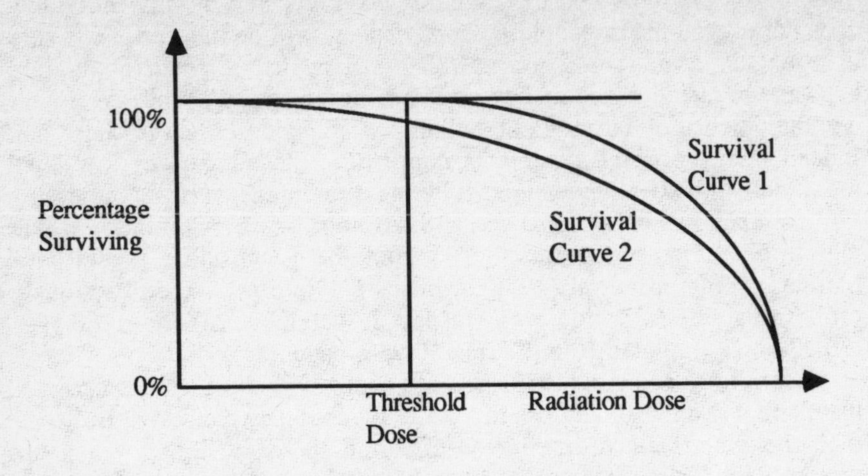

Figure 2.2 Survival Curve and Threshold Dose

radiation, no matter how little, produces a response. This result is based upon data where the smallest total dose is 20 rads (200 mGy) and these higher doses are then used to extrapolate to very low doses. It is also a laboratory experiment on human cells rather than a study of actual people. Hence, whilst the study supports the absence of a threshold dose, it does not conclusively prove it.

Considerable evidence is presented in subsequent chapters of increases in cancer having been caused by very low doses of radiation e.g. doses well within the current occupational safety limits of 5 rems (50 mSv) per year. It is evidence such as this which has led to the acceptance of the view that there is no threshold, and that all additional radiation is dangerous.

The maximum permissable dose of radiation is sometimes, rather misleadingly, referred to as the safety standard. This use of the word safe can lead to the maximum permissable dose being interpreted as the threshold dose, i.e. the largest dose for which the health effects are zero. This confusion has been fostered by the nuclear lobby who wish the industry to be seen as totally safe. For example, the deputy director (policy) of the National Radiological Protection Board (NRPB) has written that

> 'dose-equivalent limits are not scientifically determined dividing lines between life and death or between safety and danger: they represent a view of the upper level of risk workers should be asked to bear. ICRP makes this clear to those who read closely the entire Publication 26. Unfortunately, this is not the sense in which many treat the issue and both ICRP and the authorities in their underlying attitudes and statements have, in effect, encouraged the heresy',

Richings (1979). Since it is now generally accepted there is no threshold dose for radiation, any additional radiation, however small the dose, is dangerous. This means that radiation exposures within the maximum permissable dose are dangerous, which is not something the nuclear industry is keen to admit. When pressed on this point the nuclear industry will respond that the safety standards represent maxima, and actual doses are less. But if it is thought that doses of radiation ought to be limited to some lower level, why were the safety standards not set accordingly?

There has been an attempt to effectively reintroduce the concept of a threshold dose. It is argued that if the dose per person is very small then the additional risk per person is so small it can be ignored. Thus the risk from small doses need not be considered so, in effect, there is some threshold dose below which the risk can be treated as zero. For example, Fry et al (1976) have written that 'there must be some level of individual risk about which there would be a general consensus that anybody modifying his behaviour so as to attempt to avoid that risk would be regarded as mad, and summations of such risks, over no-matter how many individuals, as meaningless'. Considered from the point of view of an individual this can seem a persuasive argument. For example, if exposure to one rad (10 mGy) per year leads to an increased risk of death from cancer of only 1 chance in 10,000 per year, many people will regard this as an inconsequential risk so far as they are concerned. But from the viewpoint of society, if 20 million people are exposed to such a risk, 2,000 people will be killed each year! Such an effect cannot be dismissed as inconsequential so far as society is concerned. Hence whilst the risks to any individual may be small, the effects upon society can be significant. Therefore it is inappropriate to ignore the risks from low doses of radiation.

Whilst it is now generally accepted that there is no threshold dose below which radiation does not cause cancer, there are other adverse health effects of radiation for which a threshold does exist. There is a threshold dose for the significant permanent impairment of organ function for the testis, ovary and lens of the eye, Pochin (1983, pp. 104-106). That thresholds exist for other adverse health effects should not be confused with the lack of any threshold dose for cancer. The carcinogenic effects of radiation have no threshold, whilst there are thresholds for the significant impairment of organ function. So, when all adverse health effects are considered, there is no safe dose of radiation.

2.4 Some Reasons Why Radiation Risks May be Understated

2.4.1 Concavity

There are grounds for thinking that the true dose-response curve is concave and, at low doses, lies above a linear dose-response curve through the origin. This implies that risk estimates set using a linear dose-response curve through a high dose observation and the origin will understate the risk at low doses. There are three reasons for expecting the true dose-response curve to be concave:-

(a) *Heterogeneity.* Humans are a heterogeneous population with different susceptibility to cancer due to genetic differences and sensitivity which varies with age, sex, environmental factors etc. Baum (1973) presented an example where the population exposed to radiation consisted of a number of homogeneous sub groups each having a different dose-response curve. He showed that such a heterogeneous population had a concave dose-response curve in aggregate. The curve had a steep initial slope reflecting the response of the most sensitive subdivision of the population. At somewhat higher doses, cancer has been induced in nearly all of this subdivision, and consequently the response of this group saturates, and the less sensitive subdivisions contribute any additional cases which are induced. These less sensitive groups have a less steep dose-response curve. It has also been argued by Morgan (1975a and b, 1978, 1979b, 1980) that heterogeneous populations will have a concave dose-response curve. Bross et all (1972) have concluded that some groups are indeed more susceptible to low level radiation. Hence heterogeneity of response may be causing concavity of the dose-response curve.

(b) *Cell Sterilization and Cell Death.* A low dose of radiation may mutate a cell possibly leading to cancer, whilst a higher dose will either kill the weak cells that are likely to initiate cancer or remove their ability to divide. This feature of high doses of radiation is used therapeutically to kill off cancerous cells. Thus above some 'optimum' dose the dose-response curve will slope downwards because the dose is killing cells rather than causing cancer, Mole (1975b), Brown (1976) and Morgan (1975a and b, 1978, 1979b, 1980). Hence, as shown in figure 2.3, the dose-reponse curve is concave.

Since the safety standards at low levels of radiation have been set by extrapolation from high doses, the risks at low levels will have been understated. For example in figure 2.3 the actual response may be B whilst the response estimated using a linear dose-response curve is only C. Pochin (1978) considers the fact that the dose-response curve rises to a maximum and then declines, and concludes that 'linear extrapolation from effects of doses higher than the maximum could give an underestimate of the effects of low doses'.

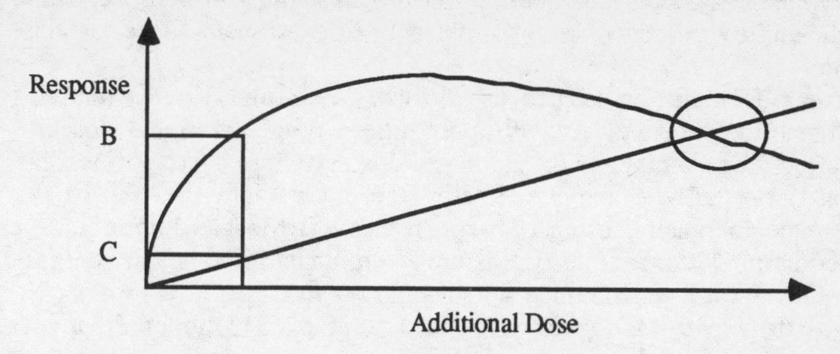

Figure 2.3 Concave Dose Response Curve When There is Cell Sterilization and Cell Killing

(c) *Follow-up Period.* Extrapolations have been made using human data where the follow-up period has been no more than 20 years. As some cancers can take over 30 years to become apparent in humans, the risks will be understated. Since at low doses cancer can take longer to develop, this bias will lead to a greater understatement of risks at low than at high doses. Hence if the estimated dose-response curve is linear, the reality is a concave function everywhere above the estimated curve. Morgan (1975a, 1978, 1979b, 1980).

2.4.2 Understatement

There are also reasons for thinking that the true dose-response curve, whatever its shape, lies above the estimated dose-response curve. This implies that, to the extent that these factors have not been fully allowed for, the calculated risks understate the true risks. There are five ways in which the estimated dose-response curve may be biased downwards:-

(a) *Reduction in Disease Resistance*. Cancers of the reticuleondothelial system (RES) are liable to damage the body's immune system. The result is that people who would have developed cancer at a later date, die of some other disease first. Kneale and Stewart (1978) used the Oxford Survey data on 10,556 children who died of cancer (and an equal number of controls). They concluded there is probably a period of several months before the diagnosis of certain childhood cancers when infection risks are abnormally high. They also estimate that, for children born between 1940 and 1950, about half the preleukaemias died of pneumonia first before cancer was diagnosed. This explains the upsurge of acute leukaemia deaths that followed the introduction of antibiotics, since the preleukaemias now survived to die of leukaemia. To the extent that preleukaemics die of other diseases due to the leukaemia damaging their immune system, the death rate from leukaemia will be understated, i.e. the entire dose-response curve lies above the estimated dose-response curve.

It has been further argued by Morgan (1978 and 1979b) that this resistance to disease effect will apply more strongly at higher doses of radiation. For example, if the estimated dose-response curve is linear the actual curve will lie above the estimated curve, with the size of the difference increasing as the dose increases. If Morgan is correct, the reduction in resistance to disease is not an argument for concavity but it is an argument for a general understatement of the risk.

(b) *Incorrect Diagnosis.* BEIR (1980, page 313) reports that post mortem studies of Japanese atomic bomb survivors indicate that lung cancer was underdiagnosed on death certificates by about one third, whilst Radford (1983) reports a figure of 45%. Hence mortality data based on Japanese death certificates seriously underestimate the risk of lung cancer. In addition, the risk of death from cancer must be distinguished from the risk of contracting cancer, with the latter being higher. Thus incorrect diagnosis will cause an underestimate of the risks of death from radiation induced cancer. Also, to the extent that non-fatal cancers are not diagnosed, there will be an understatement of the risks of contracting cancer.

(c) *Healthy Worker Effect.* When conducting an occupational study, allowance may not be made for the 'healthy worker effect'. This arises because, if a person is healthy enough to work, he or she is in a selected group which has lower mortality rates for any disease than the population in general. The general population includes people who are ill or dying, the chronically disabled etc. Najarian (1978) reports that for the U.S. most groups of working people have death rates that are only half those for the U.S. population as a whole of the same age and sex. However, some studies of radiation workers have compared them with the general population and reached the silly conclusion that being exposed to radiation improves one's health!

(d) *Non-Uniform Distribution.* Morgan (1975b) pointed out that the ICRP and NCRP standards for internal exposure to radiation assume that the radioactive material is evenly distributed throughout the body organ concerned. However Morgan has argued that this leads to a serious underestimate of the risk of bone cancer from plutonium-239 because this substance is not evenly deposited on bone surfaces. Instead the level of deposition is concentrated in areas which are more sensitive to radiation, leading to an underestimate of the true risk.

(e) *Cross Irradiation Between Organs* . Morgan (1975b) also pointed out that the ICRP and NCRP standards for internal exposure to radiation assume irradiation only from the deposits of radioactive material within the organ concerned. Where cross irradiation between organs occurs there will be an understimate of the risks.

2.5 The Concavity of the Dose-Response Curve and the Inadequacy of the Data

The shape of the dose-response curve is unknown and there is no evidence available which is conclusive. A number of authors have suggested the dose-response curve may be concave.

"There are three types of human malignancy which allow a comparison to be made of the rate of cancer induction by high and low dose irradiation: thyroid carcinoma, leukaeumia and breast cancer. In each case the evidence contradicts there being a lower risk per rad at low doses and/or low dose rates than at high doses and dose rates. Indeed, in many cases, the data favour a higher risk per rad compared to high doses." Brown (1977).

"The cancer risk estimates for exposure to low doses based on the straight line extrapolation could be somewhat lower than might be found eventually to apply, especially to susceptible subsets of the population." Radford in BEIR (1980, page 230).

"Extrapolation from high doses can lead to an underestimate of the effect of low doses." BEIR (1980, page 20).

An alternative approach to determining the nature of the dose-response curve is theoretical rather than empirical . A theory of the mechanism by which radiation causes cancer will imply a particular shape for the dose-response curve. However, since there is little agreement on which, if any, of the various theories is correct, this approach is not very helpful.

On this point Gofman (1983, page 61) has stated that he 'is appalled, on reading the Minority Report of the BEIR-III Draft Report (1979) to find scientists talking about the use of radiobiological theory in the prediction of the relationship of radiation doses to frequency of induction of human cancer. It is quite amazing to find scientists invoking radiobiological theory about a process that none of us, they included, have the ability even vaguely to describe, since that first cancer cell cannot be seen'. The final BEIR report contained the statement that "because we do not yet have an adequate theory of cancer induction it is not possible to derive a theoretical basis for dose-response data for these effects from first principles", BEIR (1980, page 21). So the main method of determining the shape of the dose-response curve is empirical. In chapters 6 to 10 some empirical evidence on the response of humans to low levels of radiation will be presented. This evidence indicates that the dose-response curve may be concave, with actual risks at low levels above the risks predicted using a linear dose-response curve.

It is sometimes stated that the connection between exposure to radiation and cancer is one of the most heavily researched health risks. However, this reflects both the difficulty of studying radiation effects and the paucity of the research on other health hazards rather than excessive research on radiation exposure, and a number of authors have pointed out the lack of adequate data on which to base radiation standards. They have called for further empirical studies, and this will be demonstrated by a number of quoatations:-

There are 'serious gaps in our knowledge of the health effects of low-dose ionizing radiation. The need for additional epidemiological study of effects at low doses is evident. At the present time, no successful systematic representation of the risks associated with exposure to ionizing radiation at low doses exists'. Landau (1974).

'The need for much more experimental and epidemiological data at low doses and for low dose-rates is apparent.' Baum (1973).

'It is very hard to estimate the risk of malignancy entailed in exposure to low doses of ionizing radiation.' Pochin (1976a).

What is needed is 'adequate support to the research programs designed to define more accurately the risks from human exposure to ionizing radiation'. Morgan (1978).

'There is a continuing need for federally sponsored research in this area (the cancer risks of low-level ionizing radiation exposure). GAO also believes that Federal research efforts can be strengthened.' Comptroller General (1981, page iv).

'It is imperative that the effects of low level radiation on humans be further researched before we implement elaborate diagnostic X-ray techniques and/or make an irrevocable commitment to pollute the environment further with hazardous radioactive material.' Bertell (1977).

'Because of the way in which they are derived, risk factors used in radiological protection must be regarded as approximations. It is essential therefore to make use of every opportunity to test the validity of the present estimates.' NRPB (1981).

In 1970 the Nobel Laureate James D. Watson said that "the amount of research now being done on the connection of cancer and radiation is totally inconsistent with proposals for wide-spread introduction of nuclear power plants into highly populated areas', quoted in Gofman and Tamplin (1979, page 117).

'Estimates of leukaemia risk for adults are scientifically reliable only within the range of observations above 20 rem or so. Any statement about this risk below 20 rem is likely to be seriously in error.' Jacobson et al (1976).

"Except for foetal exposure we have very little direct information about the possible effects on human beings of a single rad." Land (1980).

"Unfortunately, the dose-response curves observed for different cancers and species vary to such an extent that none of the parameters can be assumed known." Land (1980).

"In view of the multistage nature of the cancer process and our uncertainty as to the mechanism(s) of carcinogenesis at low doses, the actual shape(s) of the dose-response curve(s) for low level radiation carcinogenesis remain(s) to be defined." Upton (1983).

"Efforts to define the mathematical relationship between dose and effect under conditions of low-level exposure in human populations must be based on assumptions and extrapolations of unproved validity." Upton (1983).

'Good dose-response data in human populations of large enough size to provide statistically reliable risk estimates in the range of doses less than 50 rad are very rare. Such data are needed if extrapolation to lower doses is to have any precision.' Radford in BEIR (1980, page 236).

'The carcinogenic and genetic effects on populations of exposure to low doses of radiation, crucial though they are, are realms of great uncertainty. Faced with this situation, researchers in the field, understandably enough, often adopt a "professional optimisim", believing that firmer conclusions can be drawn from their results than is actually the case. The national and international committees which try to estimate radiation risks and then recommend radiation safety standards are mainly staffed by these researchers. Consequently, the statements made by these bodies are often more definite and less cautious than is justified by the precision of the data.' Barnaby (1980).

'Current data are obviously insufficient to settle the question of human low-dose effects.' Conclusions of a White House Interagency Task Force quoted by Marx (1979).

"For human populations in particular, knowledge of dose-response relationships is too limited to enable confident prediction of the shapes and slopes of the curves at low doses and low dose rates." ICRP (1977, page 7).

'Both sides in the dispute agree that there is insufficient human data to provide statistically valid risk estimates for exposures of less than about 10 rads.' Dickson (1979).

'We lack adequate information on the effects of low radiation doses in human populations.' BEIR (1980, page 28).

2.6 Absolute and Relative Risk Models

When estimating a dose-response curve, as well as specifying the mathematical form of the dose-response curve, it is also necessary to specify the way in which the additional risk varies over time. There are two alternative models of the way in which exposure to radiation causes an increase in the underlying cancer rate; the absolute and relative risk models, and at present there is insufficient empirical evidence to make a clear distinction between them.

The absolute risk model assumes that, for a given dose, the cancer risk increases by some fixed amount which does not alter as the exposed subject subsequently grows older. Figure 2.4 shows that, in the absence of any additional radiation exposure, the risk of cancer increases with age i.e. the line AJB. The precise position of the line AJB depends on the age of the subject at the time of exposure. After allowing for a latent period of OC years, figure 2.4 shows an increase in the risk of cancer of DJ. The size of this increase in the cancer risk depends on the age at exposure (see chapter 1). Hence the time profile of cancer risk becomes AJDE. The key feature of the absolute risk model is that the cancer risk in subsequent years is increased by the constant amount DJ i.e. the line DE lies everywhere the distance DJ above JB. As the underlying cancer risk rises with age, over time the additional cancers induced by the radiation exposure become a smaller and smaller proportion of the total.

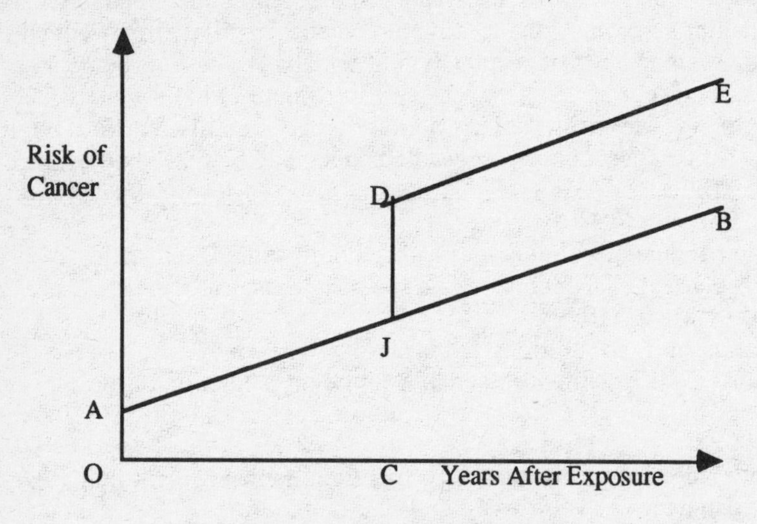

Figure 2.4 Absolute Risk Model

As well as being illustrated graphically, the absolute risk model can also be stated mathematically:- TR(t) = SR(t) + X(t), where TR(t) is the total risk of cancer t years after exposure, i.e. the line AJDE in figure 2.4, SR(t) is the spontaneous risk of cancer t years after exposure i.e. the line AJB in figure 2.4 and X(t) is the increase in the cancer risk t years after exposure caused by the additional radiation. During the latent period X(t) equals zero, and subsequently it is equal to some constant number i.e. DJ in figure 2.4.

The relative risk model assumes that, for a given dose, the increase in the cancer risk is some fixed proportion of the spontaneous age-adjusted cancer risk as the subject grows older. In figure 2.5 the spontaneous cancer risk is the line AJB, whose location depends upon the age of the subject at the time of exposure. After a latent period of OC years, the cancer risk increases by FJ, where the size of FJ depends upon the age at exposure. As the spontaneous cancer risk rises with age, and the additional radiation is assumed to produce a constant proportionate rise, the extra risk of cancer due to the radiation exposure becomes a larger and larger absolute amount. Thus in figure 2.5 the time profile of cancer risk becomes AJFG.

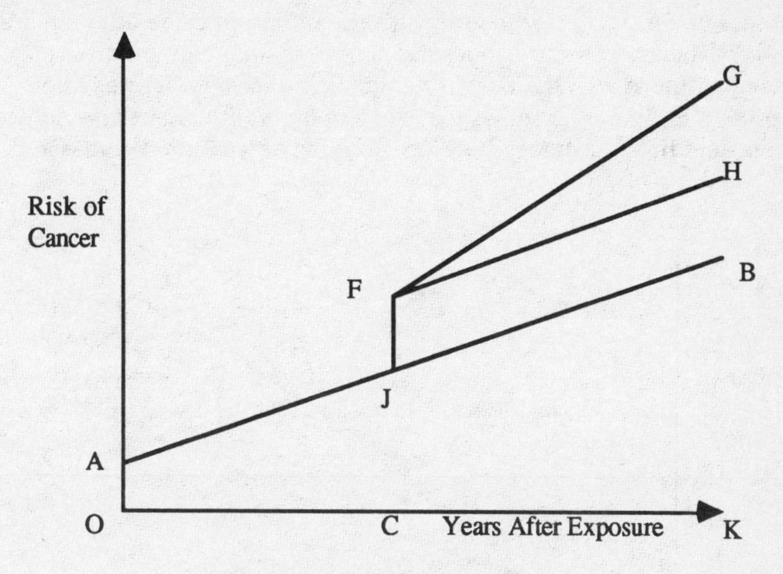

Figure 2.5 Relative Risk Model

The relative risk model may be stated mathematically as:- $TR(t) = (1 + X(t)) \times SR(t)$, where $X(t)$ is now defined as the proportionate increase in the spontaneous cancer risk t years after exposure, and the other terms are as previously defined. During the latent period $X(t)$ equals zero and subsequently it is equal to some constant. Thus OC years after the exposure, $X(t)$ is equal to FJ/CJ and also to GB/BK.

Empirical studies of the cancer risks from radiation exposure generally suffer from an inadequate follow-up period. In chapter 1 it was pointed out that the latent period for cancer may be over twenty years, whilst cancer deaths may occur up to fifty years after the expiry of the latent period, and so it is difficult to determine whether the absolute or relative risk model is applicable. All that can be estimated is the size of the initial increase in cancer risk, e.g. DJ or FJ. Having used the data to estimate the initial increase, either the relative or the absolute risk model must be assumed in order to estimate the lifetime risk. If a relative risk model is assumed, the increase in the lifetime risk is considerably larger than if an absolute risk model is used. This can be seen from figure 2.5, where the line AJFG represents the relative risk model and the line AJFH represents the absolute risk model. The area FGH is the extra lifetime risk due to the use of the relative risk model. The view of Gofman (1983, page 112) on the validity of the relative risk model is that 'the evidence strongly favours it'.

Another factor in the estimation of the increase in lifetime risk is the length of the time over which the extra risk continues. Does the increased risk continue for the rest of the subject's life or does it cease after a number of years? The answer to this question can make an important difference to estimates of the size of the risk from radiation exposure. For example, the shaded area in figure 2.6 shows the effect upon lifetime risk if the subject's pre-exposure life expectancy (OK) exceeds the expression period (CL).

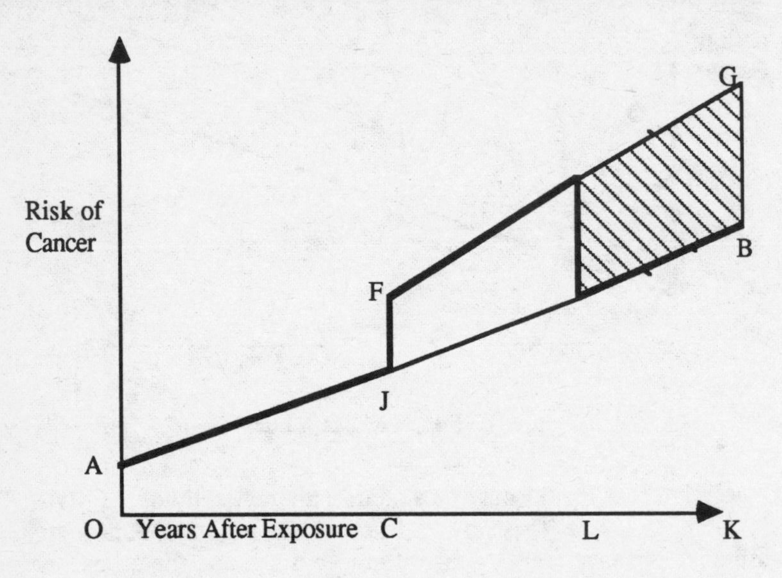

Figure 2.6 Relative Risk Model with a Limited Expression Period

2.7 Summary

Since the shape of the dose-response curve cannot be specified on theoretical grounds, its shape must either be estimated or assumed. There is a lack of adequate data on which to base such estimates, and so various official bodies have assumed the dose-response curve to be linear. The linearity assumption has considerable practical advantages. It greatly simplifies the estimation of the dose-response curve. It also makes the setting of safety standards much more straightforward, since the cancer risks of a given total dose are independent of its distribution and of the previous dose received by those concerned. However, evidence is presented in this book to suggest that the actual dose-response curve may be concave, and that the linearity assumption should be abandoned.

It is now generally accepted that there is no threshold dose i.e. there is no such thing as a safe dose of radiation and any dose of radiation is dangerous. There have been attempts to try to reintroduce the concept of a threshold dose by the back-door but these should be resisted, and statements that small doses of radiation within the exposure limits are harmless should be challenged.

Besides there being some empirical evidence to support the view that the dose-response curve is concave, there are also some theoretical reasons for expecting the curve to be concave (heterogeneity of the population, cell sterilization and death and the length of the follow-up period). There are also reasons for thinking that empirical estimates of the dose-response curve, whatever its shape, understate the true risks at every dose level (reduction in disease resistance, incorrect diagnosis, the healthy worker effect, non-uniform distribution in the organ and cross-irradiation between organs).

Finally there is the additional problem whether an absolute or a relative risk model is most appropriate. It will be shown later in chapter 5 that the choice between these models can have an important effect upon the resulting risk estimates.

3 The controversy over radiation risks

3.1 Introduction

This chapter will show that risk estimates are subject to considerable controversy, even within the official organizations which pronounce upon them. It will then be shown that governments have acted to prevent and suppress research which is critical (or likely to be critical) of the existing risk estimates. Finally the role of the NRPB, a government organization supplying advice on radiological protection, will be discussed. It will be argued that, rather than acting in an impartial manner, the NRPB has taken a partisan role on the side of the nuclear industry.

3.2 The BEIR III Saga

An authoritative report on the effects of low level radiation was produced by the Advisory Committee on the Biological Effects of Ionizing Radiations (BEIR) of the US National Academy of Sciences (NAS) in 1972. In the late 1970's it was decided to produce an updated version of the report to reflect the interim scientific advances. On 2nd May 1979 the BEIR report was released at a press conference and submitted to the EPA. However, the report was accompanied by a strongly worded minority statement objecting to the recommendations on the cancer risks from low level radiation. It then emerged that over half of the BEIR somatic effects subcommittee shared the reservations of the minority report. In consequence Dr. Philip Handler, President of the NAS, asked for the relevant section of the BEIR report to be redrafted. A revised report incorporating the views of the minority statement was published in 1980, BEIR (1980). This time the report contained a minority statement from Professor Edward Radford, Chairman of the BEIR III Committee and Chairman of the BEIR Subcommittee on Somatic Effects.

In his minority statement Radford made a number of criticisms of the BEIR report, of which Advisory Committee he was the chairman. In October 1977 the subcommittee had a vote on whether or not to use linear dose-response curves and linearity won. This vote was never rescinded, and yet the revised BEIR report used non-linear convex dose-response curves. Due to lack of time "a detailed and critical discussion by the subcommittee of the scientific basis of deciding whether one or another dose-response model was applicable to cancer risks was not undertaken"; Radford in BEIR (1980, pp.228-229). Hence the shift to non-linear dose response curves in BEIR (1980) was not based upon detailed consideration of the scientific evidence, and conflicted with a democratic decision taken by the subcommittee.

Radford points out that, by introducing non-linear dose-response curves for the somatic effects of radiation, BEIR (1980) contains an inconsistency between the conclusions of the somatic effects subcommittee and the genetic effects subcommittee. This is because the genetic effects subcommittee used a linear dose-response curve, BEIR (1980), page 321). Finally, Radford argues that much of the case for using non-linear dose-response curves is based upon animal studies, and he then goes on to criticise the use of animal studies for setting human safety standards, BEIR (1980, pp.232-233). Radford is on record in congressional testimony as being in favour of a tenfold reduction in current occupational exposure limits to radiation, Dickson (1979) and Marx (1979).

The consequence of this controversy over its recommendations is that the credibility of BEIR III has been seriously damaged. This episode highlights the controversial nature of setting safety standards for low level radiation.

3.3 Official Obstruction

The controversy over the dangers of low level radiation has polarized the scientific community into two groups. In the first group are those scientists who broadly accept the official safety standards. In the second group are those scientists who think official estimates of the risks of low level radiation are too low. A number of scientists have changed groups after becoming convinced that the risks are considerably greater than official estimates. There do not appear to have been any changes of mind in the opposite direction.

If a scientist joins the second group the consequences can be considerably to the disadvantage of his or her career. In order to carry out research into the risks of radiation a scientist needs access to the data, and also money to finance the study. Both of these essential factors can be removed or withheld from scientists who are known to think that radiation safety standards are too lax. The government and the nuclear industry, both in the U.S. and the U.K., are in a position to prevent such a scientist receiving funding. Much of the data referred to is held by government agencies including various nuclear organizations, e.g. the health records of nuclear workers. The government can also create difficulties for researchers who wish to publicise their findings. Even when safety standards have been set, there can be difficulties for safety officers in ensuring that safety procedures are followed. These problems will be considered in the following sections.

3.3.1 Access to Data

This section contains a number of examples which typify how governments use their power to prevent or obstruct researchers from obtaining data.

(a) Hanford The U.S. Department of Energy heard that Dr. Samuel Milham had discovered increased cancer risks among workers at the Hanford reprocessing plant. Funded by the Department of Energy, Dr. Thomas Mancuso had been collecting data on the Hanford workers since 1964. In order to discredit any public statement by Milham, Mancuso was asked to issue a press release "denying any increased cancer risk". However, since Mancuso had not completed his analysis, he refused to issue such a press release. According to Bross (1981, page 218) the result was that the U.S. government concluded Mancuso was a potentially dangerous man. As custodian of extremely sensitive data on the health and radiation records of Hanford workers, Mancuso could not be trusted to suppress any positive results; he might produce a report critical of the current safety standards. Therefore Mancuso's study had to be stopped.

The Hanford data was stored on computer tapes. An ERDA officer contacted the computer centre at ORNL and, over the objections of Dr. Mancuso, confiscated most of his data. Fortunately, Dr. Mancuso had a copy of his data at the University of Pittsburgh. The ERDA then tried to confiscate all Dr. Mancuso's data, but this was successfully resisted by Dr. Mancuso and the University of Pittsburgh, Bertell (1985, pp.91-92). The ERDA then circulated critiques of the Hanford study in the U.S.A. and Europe without making a copy available to Dr. Mancuso. Indeed, he was only able to obtain a copy of these critiques of his work by invoking his rights under the Freedom of Information Act.

The Department of Energy called in six reviewers to decide whether or not to continue support for Mancuso's study at the University of Pittsburg. Four of the six reviewers were uncritical and recommended that support be continued. One reviewer was critical, but did not suggest taking the study away from Mancuso. Only the sixth reviewer recommended the study be taken away from Mancuso. On the basis of these six reviews, but mentioning only the two that were critical, a member of the Department of Energy staff wrote a memo in January 1976 recommending termination of the Mancuso contract. The contract was subsequently transferred by the Department of Energy to the Oak Ridge National Laboratories (ORNL), where it was under the administrative control of Dr. Sydney Marks, the very same Department of Energy employee who had written the termination memo prior to leaving. Subsequently, Mancuso, Stewart and Kneale obtained some financial support from a privately funded environmental action group, the Environmental Policy Centre, Marx (1979). The results of the Hanford study are discussed below in chapter 8.

The ERDA tried to conceal the fact that they had received four uncritical reviews of the Hanford study. During congressional hearings, Congressman Tim Lee Carter asked whether the two negative reviews were the only ones received by the ERDA prior to the decision to terminate Dr. Mancuso's funding. Dr. Liverman of the ERDA, who was giving evidence under oath, replied that this was true. Carter then produced copies of the four uncritical reviews which had been received by the ERDA at the same time as the two negative reviews, Bertell (1985, page 92).

In his final report in July 1977 Mancuso discussed the ending of his contract and wrote that 'this decision to terminate the project at the University of Pittsburgh, in the light of the positive findings of a definite relationship between work exposure to ionizing radiation and cancer is, in my opinion, not in the best interest of science', quoted by Grossman (1980, page 95).

(b) Portsmouth When Dr. Najarian wished to study the effects of low level radiation on the dockyard workers at Portsmouth, U.S.A., he was refused cooperation by the U.S. Navy. Even though the Navy had both the health records on its workers and records of the radiation doses to which they had been subject, it repeatedly refused to supply them to Najarian. This could easily have stopped Najarian, but he managed to carry out a limited study with the help of five reporters from the Boston Globe working full time for six weeks interviewing the next-of-kin; Najarian's results are discussed below in chapter 8. Subsequently the local senator got the U.S. House Health and Environment sub-committee to hold hearings into the effects of low level radiation at the Portsmouth dockyard. The committee chairman pressed for and won an assurance from the navy

that there would be a full investigation of the health effects of low level radiation, Torrey (1979).

(c) Sternglass When he suspected that fallout from nuclear tests had led to high cancer rates, Sternglass (1981, pages 45 and 51) was refused data on childhood cancer rates in the Troy area of New York state by the New York State Health Department. A further request for data was also refused. Later, when he wished to study the effects of having nuclear power stations in the locality on local infant mortality rates Sternglass (1981, pp. 124-125) was unable to obtain any funds to finance such studies. Fortunately he found a group of concerned students who were willing to help him. His results are presented below in chapter 9. Sternglass (1981, page 271) reports that the budgets of public health agencies were cut to stop the publication of the detailed annual health statistics which would have allowed investigations, such as his own, to be continued. Sternglass claimed that the fragmentary summaries of data which replaced the detailed reports in 1970 were a very inadequate substitute. In 1974 the Nixon administration cut the budget of the EPA to force an end to the publication of 'Radiation Health Data and Reports'. In 1979 the previously required monitoring of strontium 90 was ended.

(d) Mururoa Since 1966 the French government have been testing their nuclear weapons at the island of Mururoa in Polynesia. They used to publish regular health statistics but this ceased one month before the testing began in 1966. The publication of environmental studies also ceased. The native Polynesians are concerned about the lack of any reports on the levels of radioactivity since the testing began and suspect that cancer rates have risen, but are unable to get any data, Mackenzie (1982).

Bertell (1985, page 99) states that persons requesting public health statistics for Mururoa after June 1966 were reported to the secret police and, according to Bertell (1985, page 166), 'it is not unusual for Polynesian critics of the French testing to be imprisoned in France'. In 1982 France refused to cooperate with a WHO study of cancers among Pacific Islanders, Bertell (1985, page 100). In June 1972 the UN Conference on the Human Environment formally condemned the French tests and criticised the inadequacy of French reports to the UN, Bertell (1985, page 101).

While trying to disrupt tests at Mururoa in 1973 David McTaggart, the founder and chairman of Greenpeace, was beaten up by the French navy. As a result, he required an emergency operation to save the sight in one eye, Sunday Times Insight Team (1986, page 100).

(e) South Pacific In the summer of 1984 the Australian government appointed a Royal Commission to enquire into the British nuclear tests in Australia in the 1950's. The president of this commission, Mr. Justice James McClelland questioned the level of co-operation by the British government. In particular, he accused the British government of

not supplying the commission with the relevant documents, Arends (1985). His explanation for this lack of co-operation by the British government was that full disclosure might have led to compensation claims from people injured by the nuclear tests.

Apparently, because they fear what may be said by the Australian Royal Commission, the British MoD funded an official history of the British nuclear tests in Australia. Ironically, the MoD paid for this study out of money saved from the budget for supplying information and facilities to the Royal Commission. The MoD can censor this official history and, according to Denzel Davies M.P., 'it looks like an attempt to blunt events which may be embarrassing to past Tory governments', Leigh et al (1985).

Some of the British tests took place at Maralinga, in South Australia. After one test in 1957 an Aboriginal family claimed to have camped in the bomb crater for two days. An Australian serviceman has given evidence to the Royal Commission that 200 soldiers were assembled and told by a British officer that anyone who informed the media of this incident would be tried for treason. The men were also reminded that, under the Secrecy Act, they could be shot by firing squad or jailed for thirty years if they talked, Milliken (1984) and Smith (1985, page 125). The effects of British nuclear bomb tests in Australia are discussed further in chapter 7.

(f) Japan When the atom bomb was dropped on Hiroshima on 6th August 1945, twenty Americans were being held in the city as prisoners of war. Eighteen died instantly and the other two died shortly afterwards. This event was classified by the U.S. on grounds of 'national security', and the mother of one of the men involved was told that her son died on 28th July 1945 of wounds suffered when his plane was shot down over Honshu Island. This deception only came to light in 1970 when a Red Cross report was declassified, Bertell (1985, page 139).

The paramount source of data for estimating the risks from exposure to radiation is the Japanese bomb data, which has been collected by the Atomic Bomb Casuality Commission (ABCC). Dr. Rosalie Bertell (1982) has claimed

> 'that the basic ABCC data i.e. records of the radiation doses and the subsequent health problems of atomic bomb survivors is classified and not available to scientists who request access to it for legitimate research purposes. The atomic bomb findings have been challenged by Dr. Joseph Rotblat, Dr. Alice Stewart and myself. We have asked to test our hypotheses against this data base, expecting government cooperation to resolve the discrepancies between the atomic findings and the findings of researchers using populations such as nuclear workers and people exposed to medical X-rays. All requests for access to information have been refused. In a letter Kelly Clifton (the permanent director and chief of researcher of the Radiation Effects Research Foundation (RERF) wrote to the

> Department of Energy after my request, he stated: "In view of Dr. Bertell's bias, it would seem inappropriate for RERF to establish any formal relationship with her". In 1978, when I was in Japan, Dr. Issei Nishimori, an internationally known scholar at the University of Nagasaki, was refused the RERF data on original atomic bomb radiation doses for individuals he had collected health data on for a number of years. I was told by Dr. Stuart Finch, then director of RERF, that Dr. Issei Nishimori "might be biased against radiation"'.

A crucial aspect of any epidemiological study of radiation is the estimation of the quantities of radiation to which people were exposed. The radiation doses from the two Japanese bombs were estimated in 1965 by John Auxier of ORNL in Tennessee. His dose estimates are called T65D, which stands for tentative 1965 dose estimates. Some of his work was classified because it was concerned with the details of the two bombs. The T65D dose figures were generally accepted, and all subsequent official estimates of risk have been largely based upon the Japanese results. Thus the T65D figures have played a key role in risk estimation. Because Auxier's calculations were based upon classified material and never described in detail, it was very difficult for other researchers to check his calculations or to conduct their own independent analysis. According to Auxier 'we knew at the time that the answer we had (in 1965) wasn't good enough but we had an answer, and the funding dried up', Marshall (1981c). So, despite Auxier having described his estimates as tenative, no serious attempt was made to check these crucial dose estimates for another 12 years.

In 1976 and 1977 however, further details of the basic data and of Auxier's calculations were made available, and this finally permitted other researchers to examine his results. The preliminary findings of these researchers did not become public until about August 1980, although they were unofficially known about by some scientists in 1977 or 1978, Radford (1981b). The major dose revisions which resulted from this work, and which will result in a considerable upward revision of the risk estimates, are discussed in chapter 5. The reassessment of the Japanese doses was funded by the Office of Target and Damage Assessment of the Defence Nuclear Agency. David Auton said that his agency was funding this research because 'nobody else was interested', that 'this work is of marginal interest to us and we really can't afford to spend very much money studying civil effects' and that it would make more sense for the Department of Energy or the Nuclear Regulatory Commission to pay for this work, Marshall (1981a).

In view of the crucial nature of the T65D figures and their tentative nature, it is surprising both that the basic data was kept classified and that no study was initiated by the responsible agencies to verify Auxier's results. Getting funding for this work from the obvious sources may have been made more difficult because of the view of Auxier, an influential person in this area, that there was no need for an independent study of any apparent discrepancies in his results, Marshall (1981a).

(g) Non-collection Not only may access to data be limited, but the authorities may choose not to collect useful data. Data on human exposure to plutonium are very limited (see chapter 4), and an obvious source of such data is occupational exposure to small amounts of plutonium. However, this potential source of valuable data has been neglected. Voelz (1975) has written that

> 'it would be nice to be able to report that the long-term studies on plutonium workers have been practiced faithfully throughout the industry. Unfortunately, the follow-up of workers following termination of their employment in plutonium work has been limited to a few special situations'.

Similarly, the US government chose not to collect any data on US soldiers who were in Japan shortly after the end of World War II. In the early 1980's there was pressure from the soldiers concerned for an epidemiological study. In a cynical exercise designed to resist this pressure, the Pentagon asked the National Research Council to produce a scientific report on the matter. In a sloppy report designed to destroy the case for a study, they duly concluded that no study was needed, Smith (1983). For example, the study estimated the expected number of multiple myelomas among a group of civilians in a *35* year period and compared this with the observed number of cases for a *five* year period amongst the US soldiers who had been in Japan, Bertell (1985, page 163). Clearly, the cancer rate amongst the veterans would have to have been increased by some seven times if any excess cancers were to be found by such a study.

The irradiation of some of the Marshall Islands by a nuclear bomb test at Bikini is described in chapter 7. Whilst there has been a follow-up of those people who were most accutely affected by the radiation, there has never been a comprehensive epidemiological study of the local population. A newly appointed head doctor of Brookhaven National Laboratory resigned because the US refused to conduct an epidemiological study of the Marshall Islanders, Alcalay (1980). This failure to carry out such a study is strange since the survivors of the Japanese bombs have been extensively studied by the US to learn about the effects of radiation. Perhaps it is because the Japanese were the enemy and could not sue for compensation, whilst the Marshall Islanders were the victims of a US nuclear accident and could sue for compensation. This may also explain why the US government has failed to conduct an epidemiological study of the population living downwind of the US nuclear weapons testing site in Nevada, (see chapter 7).

(h) Data Falsification In addition to not collecting data, the authorities may falsify the data. For example, a former US army medic,

Van R. Brandon, has said that he was ordered to enter false data to hide the fact that soldiers at four atomic tests in 1956 and 1957 at the Yucca Flat test site were exposed to dangerously high levels of radiation. In February 1982 Brandon told the New York Times that

> 'we were instructed to keep two sets of books. One set was to show that no one received an exposure above the approved dosimeter reading. The other set of books was to show what the actual reading was. That set of books was brought in in a locked briefcase attached to a man's wrist by a set of handcuffs every morning',

quoted from Saffer et al (1982, page 309).

In 1984 a survey was conducted of members of the public living near Windscale(Sellafield). More local people considered BNFL and government bodies to be an unreliable source of imformation than considered local and quality national newspapers to be unreliable, Macgill (1986). This suggests that the UK public now doubts the truthfulness of official statements on the radiation risks connected with Windscale (Sellafield).

Doug Rickard's job was to issue dosimeters to men going into dangerous areas at Maralinga in Australia. Each dosimeter had to be charged by battery before being issued, but many of the batteries were flat and so the dosimeters did not work. In consequence, Rickard just made up the doses, Smith (1985, pp. 131-132).

3.3.2 Access to Money and Jobs

As well as controlling access to data, governments also provide jobs for researchers and money for research projects. There are various examples of these being withdrawn from opponents of the official line on radiation risks.

During the Sizewell B inquiry Robin Grove-White, director of the Council for the Protection of Rural England, said that

> 'in at least three instances, expert consultants who could have contributed authoritatively to the economic case against Sizewell B have had to decline to appear for us, for fear of being penalised professionally by the Central Electricity Generating Board (CEGB) when the inquiry is over',

Financial Times (1983b).

Professor Bross was in charge of a study at the Roswell Park Memorial Institute in the U.S.A. into the effects of low level radiation. However, the funds for this study were withdrawn by the National Cancer Institute, the study cancelled and the staff employed sacked, Bross (1981, pp.172-173).

In 1955 Dr. Alice Stewart set out to study the reasons for the increase in the childhood leukaemia rate since the Second World War. The Medical

Research Council turned down the project, but she financed it herself, Bunyard (1981, page 116). Her results are discussed below in chapter 6.

In Shutdown (1979, page 64) Dr. John Gofman testified that, after publishing a paper stating that radiation produces twenty times as many cancers as had been thought, he lost his grant of $250,000 per year for his cancer chromosome work. Gofman stated that, in his view, the two events were directly connected. Gofman's co-author, Dr. Arthur Tamplin, has alleged that shortly after criticisng nuclear safety standards he had 10 of his 11 staff taken away from him by the AEC. Tamplin also pointed out that the year after his criticisms was the first one in his 7 years of employment that he did not receive an annual salary increase, Boffey (1970).

Dr. Rosalie Bertell, who has conducted a number of studies on the dangers of low level radiation which are discussed in chapter 6, is not only a scientist but is also a nun. She was employed at the Roswell Park National Cancer Research Institute in Buffalo, U.S.A. In 1973 she spoke on the health effects at a public hearing on the siting of a nuclear plant in the area in which she lived. The reaction of her employers was to require her to clear all press interviews with her superiors. Hostile articles about her appeared, her mail was opened and her office ransacked. Finally, after being told what she might and might not say at a Senate hearing, before whom she had been asked to testify, she resigned. She has later written that 'I was little prepared for the military-like confrontation with the nuclear industry which my research produced', Bertell (1978). She now continues her research funded by the Ministry of Concern for Public Health which was set up by a group of Jesuits, Wardle (1983).

Radford (1981b) reports that during the controversy over the risks from exposure to radiation he was told by a highly respected scientist working in the U.S. government's laboratories, that although he agreed with Radford's position, he was under orders not to discuss any of the issues in public nor to allow his name to be associated with a particular position in the controversy. Hence this scientist was effectively prevented from participating in the debate over the risks from radiation.

Dr. Carl Johnson carried out a study of the cancer effects of the Rocky Flats nuclear plant in Colorado on the local population, and his study is presented in chapter 7. Tucker (1983d) reports that after conducting this study, Dr. Johnson was abruptly dismissed from his post as Medical Officer of Health for Jefferson County, Colorado. Tucker concludes that nobody in Jefferson County Public Health Department is likely to raise this issue again.

Before the Three Mile Island plant went into operation Dr. Chauncey Kepford of the Pennsylvania State University had warned local residents that the radiation doses to the local population would be much larger than calculated by the Nuclear Regulatory Commission. Because of his efforts to assist local citizens prepare a case against the Three Mile Island plant he was sacked by Pennsylvania State University, something he was able to prove in court when he sued the university for damages, (Sternglass, 1981, page 149).

At the time of the Three Mile Island accident Dr. Gordon Macleod was the Secretary of Health of the Commonwealth of Pennsylvania. However, on 9th October 1979 the Governor of Pennsylvania fired him. Sternglass (1981, page 264) reports that Macleod is convinced he was sacked because he wished to find out the truth about the health effects of the Three Mile Island accident.

In 1979 Dr. Barry Matthews of the Agricultural Research Council's soil survey team was sampling the soil from the Windscale (Sellafield) and Ravenglass areas. He came across an area of unusually high radioactivity (up to 100 times the natural background level) around the outlet of the Drigg stream onto the holiday beach at Seascale. (Drigg is the burial site for solid waste from Windscale (Sellafield) with a low level of radioactivity). He also discovered that the geological maps of the area were wrong. It was not composed solely of boulder clay which would help contain radioactive material. It also contained some sandy drift soils through which radioactive material might escape.

He wrote to British Nuclear Fuels Ltd., (BNFL) and was assured by their safety manager that everything was carefully monitored and was within authorized limits and procedures. So he wrote to his M.P., Lord Whitelaw, who obtained a similar bland assurance from Mr. Heseltine, the Environment Secretary. Dr. Matthews casually warned a family who were visiting the beach that it was an unsuitable area for children to play because of the radioactive contamination. For his actions Dr. Matthews was sacked in June, 1981 after 20 years service. Subsequently it was decided he had been wrongfully dismissed and he was awarded £6,500 damages. However, he was not given his job back and he remained unemployed for at least another 18 months, Tucker (1983c).

Partly as a consequence of governmental unwillingness to provide data and money to outside organizations, much of the research on the dangers of radiation is done in government research establishments and not in the universities. In the U.S.A. much of the experimental research into the biological effects of ionizing radiation has been done in government-run laboratories, e.g. Brookhaven, Argonne and Los Alamos. But these same research laboratories were also responsible for the development of nuclear energy and so have a vested interest in the acceptance of nuclear technology, Radford (1981a). This raises the issue of their suitability to also conduct the research into the dangers of this technology.

3.3.3 Publicity

Bross (1981, page 163) points out that, whilst there are disincentives for research showing that low level radiation is more dangerous than currently thought, there are incentives for demonstrating the opposite. These include publication, increased funding for research projects, appointment to prestigious committees, etc. The nuclear industry also spends very large sums of money on promoting the view that it is a safe industry, Gofman (1976b). If a study which questions the accepted levels of radiation risks is actually carried out, there are various ways in which the publicity given to such a study can be limited and prejudiced. Academic journals may refuse to publish such studies, or may do so only with official disclaimers as to the validity of the article, and include hostile reviews, Bross (1981, page 166).

According to Shutdown (1979, page 161) and Sternglass (1981, pp.211-212) the U.S. AEC suppressed an unknown number of studies on the health effects of fallout. In support of this view Lapp (1979, page 18) states that the AEC "was not averse to classifying biological and medical reports (on the effects of nuclear radiation) so they would not see the light of day". The AEC also displayed an arrogance with respect to admitting that radiation had harmful effects, and even went so far as to suppress reports of the genetic effects of radiation.

In the 1950's Congressman Chet Holifield, who went on to become chairman of the Congressional Joint Committee on Atomic Energy, said

> 'I believe from our hearings that the AEC approach to the hazards from bomb test fallout seems to add up to a party line - 'play it down'. As a custodian of official information, the AEC has an urgent responsibility to communicate the facts to the public. Yet time after time there has been a long delay in issuance of the facts, and often times the facts have to be dragged out of the agency by the Congress. Certainly it took our investigation to enable some of the Commission's own experts to break through the party line on fallout',

quoted in Grossman (1980, page 94).

In August 1980 the Congressional Committee which investigated the health complaints of people living downwind of the Nevada test site concluded 'that the AEC's desire to secure the nuclear weapon testing programme took precedence over the Commission's responsibility to protect the American public's health and welfare', quoted from Bertell (1985, page 201). In order to gain public acceptance of atmospheric bomb testing in Nevada (see chapter 7), President Dwight D. Eisenhower declared the policy of the US government to be 'keep the public confused', quoted from Bertell (1985, page 54) and Smith (1985, page 143). According to Bertell (1985, page 96), such confusion encourages reliance by the public on the status quo decisions made by prestigious government experts. It is the view of Dr. Rosalie Bertell that 'the Atomic Energy Commission lied about the hazards from the dumping of nuclear waste in the oceans by the US', Bertell (1985, page 303). A young man accidentally drove through the fallout area of a US nuclear bomb test, sustaining radiation burns from which he died. Publicity of this accident was suppressed, Bertell (1985, page 201).

Not only does the US government and its agencies try to confuse and mislead the public, they also influence legislators to prevent specific reference to the health dangers of low level radiation. In a letter written in September 1981, the General Counsel of the US Department of State, William H. Taft, tried to persuade Congressman G.V. Montgomery, Chairman of the Committee on Veteran Affiars, to delete assistance to veterans exposed to nuclear weapons testing from a bill then before Congress. Taft wrote that

> 'section 3 of the Senate passed bill creates the unmistakable impression that exposure to low level ionizing radiation is a significant health hazard when available scientific and medical evidence simply does not support that contention. This mistaken impression has the potential to be seriously damaging to every aspect of the Department of Defence's nuclear weapons and nuclear propulsion programs',

quoted from Bertell (1985, pp.68-69). The bill was subsequently modified to delete specific reference to veterans exposed to nuclear weapons testing. This letter from Taft reveals the government's fear of the public realizing that the production and deployment of nuclear weapons and the use of nuclear powered submarines results in low level radiation exposure, and that this exposure is dangerous.

Publication of an early article by Karl Morgan was delayed for almost a year in 1945 because he argued that permissible levels of occupational exposure to radiation should be much lower, Morgan (1975b).

In 1955, two scientists from Colorado University, Ray Lanier and Theodore Puck, challenged a claim by the American AEC that nuclear weapons tests posed an insignificant health hazard. The Governor of Colorado, Edwin C. Johnson, responded by saying that Lanier and Puck 'should be arrested' for their views, Smith (1985, page 145),

A study by Dr. Robert Pendleton on fallout in milk in 1962 was suppressed, and Pendleton said that funds for his studies of radon gas were cut off as a reprisal for his having conducted the study. Also a study by Edward Weiss in 1965 linking a sharp rise in leukaemia in school children to the Nevada nuclear tests was suppressed by the AEC. In squashing his study, the AEC told Weiss that publication of his study 'will pose potential problems to the Commission; e.g. adverse public reaction, lawsuits, and jeopardizing the programs at the Nevada Test Site'.

In the opinion of Tucker (1986), only about 10% of the expected number of UK government research papers on radiation risks have been published. He argued that publication of some 90% of such research reports has been blocked or delayed.

Whilst an article by Sternglass on the connection between fallout and infant mortality was published in an academic journal in April 1969, the journals' managing editor later informed Sternglass (1981, page 97) that both before and after publication of the article the editor had received pressure from Washington not to publish. A study of the effects of fallout on infant and adult mortality by Lester Lave, Samuel Leinhardt and Martin Kaye in 1971 was finally accepted for publication in an academic journal. However, just before publication Sternglass (1981, page 138) reports that the editors received objections from highly placed government officials, and the plates were destroyed. The article has never been published.

In 1969 when Sternglass was invited to appear on the NBC TV show 'Today' to discuss the link between fallout and infant mortality, the AEC called the producer and urged him not to invite Sternglass (1981, page 115). In 1979 Sternglass was scheduled to appear on the ABC TV show 'Good Morning America' the day after the publication of the Kemeny Report, the official investigation into the Three Mile Island accident. His ticket to New York had been paid for, a hotel room had been reserved and a car arranged to take him to the TV studio. But just a few hours before he was due to leave for New York he was told by ABC that his interview was cancelled, Sternglass (1981, page 249).

After Sternglass had reported evidence showing that the Three Mile Island accident had caused a rise in local infant mortality, the local newspaper *The Harrisburg Patriot* carried an editorial on the matter. This claimed Sternglass was 'inept at gathering statistics, or worse, he simply fabricated them to fit his conclusion. For a scientist to present grossly inaccurate data is inexcusable. But to fit the method of analysis to a conclusion makes the scientist's motives suspect', Sternglass (1981, page 252). Spokesmen for the nuclear industry all over the world then quoted this editorial in an attempt to discredit Sternglass' analysis of the health effects of the Three Mile Island accident. For example, an editorial by the New York Times on this matter was entitled 'Nuclear Fabulists'. Sternglass (1981, page 256).

Previously, in 1970, Sternglass had presented data purporting to show that the Dresden reactor had increased local infant mortality, and the AEC had hired a professional 'truth squad' to appear during or after radio and TV broadcasts featuring Sternglass. At a Congressional hearing a female Health, Education and Welfare (HEW) bureaucrat physically assaulted Sternglass. She was overwhelmed and subsequently hospitalised, Torrey (1980b).

The ultimate in government repression of information was the secrecy over the world's most serious nuclear accident, which occurred in the Russian Urals in 1957-58. It is clear from recently released U.S. government documents that they knew of this accident by 1958, and yet it was not publically disclosed until Dr. Medvedev, a Russian biochemist, did so in 1976. The accident is discussed in detail in chapter 10.

Scientists who are known to oppose current safety standards because they claim them to be too lax may not be permitted to speak at academic conferences. Bross (1981, page 216) reports that Dr. Karl Morgan, the father of health physics, was stopped from presenting his findings to an international scientific meeting just before he was to speak. Shutdown (1979, page 161) states that in 1971 Dr. Morgan travelled to Germany to present a paper on radiation health only to find that ORNL had wired the German authorities instructing them to destroy all 200 copies of the paper before they could be given out.

Dr. Caldicott (1980, page 87) was invited to speak to the nuclear submarine workers at the Portsmouth navel dockyard in the U.S., but only four men appeared; the workers had been threatened by the U.S. navy with the loss of their jobs if they attended the meeting.

Major reviews of the scientific evidence on the dangers of low level radiation may just ignore studies critical of the current standards, Bross (1981, page 159).

According to Bross (1981, page 168) there are a number of scientists, such as Dr. Charles Land, who make a point of criticising studies showing that the current safety standards are too lax. Bross refers to such scientists as 'professional hatchet men' and says of Land 'if there is any positive report that he has not attacked, then it has not come to my attention (or his)'. It is possible to criticise almost any study along the lines that it does not prove causality, it has not controlled for all possible factors, the sample size is not large enough, an alternative statistical technique could have been used, etc., see chapter 4.

Dr. Rosalie Bertell (1982) reports that when she and Dr. Irwin Bross published results using the Tri-State Leukaemia Survey (see chapter 6), the US government obtained copies of the data and hired Michael Ginevan of the Argonne National Weapons Laboratory to re-analyze the data. She implies that the purpose was to disprove her results.

Such practices also occur in the UK, and Peter Taylor has written that "it is our considered opinion that in Britain dissenting scientists are subject to a mixture of ostracism and depreciating pronouncements which, for whatever motivation, severely restrict their status in decision making" Taylor (1982, page 9).

An example of the low level of debate to which some scientists have descended occurred at an academic conference in California in 1971.

During discussion of a paper presented by Professor Sternglass, J.R. Totter of the AEC said:

> 'I have never before commented on Dr. Sternglass' presentation because I felt that he had so obviously and flagrantly misused data that his work should not be dignified by serious scientific comment. As statisticians or experimental biologists you know that, even in very well controlled experiments, the data never fit the hypothesis being tested so beautifully as the data Dr. Sternglass has presented. I believe, therefore, that he has convicted himself of selection of data to fit his hypothesis'.

Sternglass (1972). In other words Sternglass had so clearly demonstrated his case that low level radiation is dangerous that Totter did not believe it!

In the view of an American investigative reporter there has been widespread supression of unpleasant information on nuclear power. "For many years I've been doing investigative journalism, working for a large New York area daily newspaper and on radio and television. I also teach investigative reporting at the state university level. I've never come across an issue subjected to as extensive a cover up as nuclear power." Grossman (1980, page xvi).

Not only do governments devote considerable resources to government the power to impose stiff fines and jail sentences on those who speak out against uranium mining, Bertell (1985, page 275).

3.3.4 Observance of Safety Standards

Not only is there pressure on researchers studying the risks from radiation exposure and those setting standards, but there is also pressure on health physicists and safety officers not fully to observe the standards. Morgan (1975b) wrote of his fellow health physicists that "there were times when some of my associates were demoted or lost their jobs because they refused to yield to pressures to lower our standards or compromise for unsafe conditions". An example of such action in respect of a nuclear scientist occurred at the Atomic Weapons Research Establishment, Aldermaston.

Trevor Brown was a development manager at the Atomic Weapons Research Establishment (AWRE), Aldermaston from 1961. The AWRE conducts research into the design of nuclear weapons. From the 1950's it has machined plutonium metal billets supplied by Windscale (Sellafield) into the particular shapes required for assembly into nuclear warheads. It also uses plutonium to make fuel elements for use in the Fast Reactor Project. Machining plutonium generates dust that escapes into the atmosphere. Senior members of the UKAEA were concerned about the AWRE's poor safety record and transferred Trevor Brown from Capenhurst to the AWRE. At Capenhurst Brown had been production manager of the chemical half of the large nuclear processing factory and had gained a reputation for overcoming technical and safety problems which had caused problems at Aldermaston and Harwell. From his arrival at the AWRE in 1961, Brown persistently expressed the view to his superiors that safety at the establishment could be improved by contact with outside safety experts, better safety training and better monitoring of atmospheric contamination. These requests were turned down on the grounds that the AWRE was a secret defence establishment thereby precluding outside contacts, that better training was unnecessary, and that better atmospheric monitoring was an unwarranted extravagence.

In 1977 Brown at last obtained better atmospheric monitoring and this revealed that contamination levels were 20 to 50 times higher than had previously been thought. Brown was also given permission to consult an outside expert at the NRPB. This expert pointed out that the AWRE was not following the recommendations of the ICRP on the control of plutonium which are observed by other British establishments handling plutonium. Subsequent investigations indicated that workers had been breathing in plutonium dust during the previous 25 years, and in 1978 all radioactive work at the AWRE ceased. This led to the Pochin inquiry into radiological control at the AWRE which criticised safety procedures at the establishment, Pochin (1978b).

In 1971, with the full consent of his superiors, Brown had become a local councillor. In 1976 his constituents asked him to approach the local MP (Michael McNair-Wilson) to obtain better safety arrangements at the AWRE. Just two years before the Pochin inquiry criticised safety procedures, McNair-Wilson was assured that 'safety standards were of the highest'. In 1980 Brown was interviewed on this point in a British Broadcasting Corporation (BBC) TV broadcast. This interview took place after Brown had received legal advice from the County Solicitor of Berkshire County Council that the only objection raised by the AWRE to his appearing (that contempt of court might be involved) was irrelevant. As a consequence of this interview Brown was severely reprimanded and "retired" early. Brown is now taking his case to the European Commission on Human Rights on the grounds that when he was interviewed on TV he was acting as an elected representative, and dismissing him from his employment as a professional officer was an infringement of his political freedom, Dalyell (1982).

3.4 Karen Silkwood

In August 1972, at the age of 26, Karen Silkwood began work in the metallography laboratory of the Kerr-McGee Nuclear Corporation Plant on the Cimarron River near Crescent, Oklahoma, Rashke (1982). She performed quality checks on the plutonium pellets produced by the plant. Having joined the Oil, Chemical and Atomic Workers Union, she took part in a strike which began in November 1972 and continued for ten weeks. In August 1974 she was elected to a post on the three person union committee to bargain with Kerr-McGee over new conditions of employment. She was allocated responsibility for health and safety, and began to take a serious interest in these matters. Throughout August and September she followed contamination incidents, asked questions, interviewed workers during her work breaks and recorded her observations in a notebook. Despite her keen interest in safety matters, it was only in September 1974 during a meeting in Washington with her union's vice president that Karen learned that plutonium causes cancer. Even though she had been working with plutonium pellets for two years, Kerr-McGee had never informed her of this fact. At this meeting in Washington Karen said the fuel rod quality assurance records at the Cimarron plant were being doctored. If this could be proved it would have serious consequences for Kerr-McGee, and she agreed to obtain hard evidence.

In October 1974 Karen worked hard to find documentary evidence against Kerr-McGee and she found that someone was touching up the X-rays of fuel rod welds and manipulating the data so that fuel rods would pass inspection. The contract negotiations with Kerr-McGee were due to begin on 6th November, whilst on 13th November Karen was scheduled to supply her documentary evidence against Kerr-McGee to her union and to the New York Times. On 5th November 1974 she found that her working coveralls had a level of radioactive contamination forty times the AEC maximum permitted level, and parts of her body had twenty times the permitted dose. No explanation for this contamination has ever been found. On arrival at work on 7th November she was found to have levels of radioactive contamination in her right nostril that were ninety times the maximum permitted level. Her apartment was found to be highly radioactive, and some food in her refridgerator was 800 times over the AEC maximum. Kerr-McGee decontaminated her apartment and removed many of her personal possessions for burial at a radioactive waste dump. Karen believed that someone had walked into her unlocked apartment and contaminated her food.

On 13th November Karen was due to meet an official of her union and a New York Times reporter in a hotel in Oklahoma City. She had with her a folder of documentary evidence of falsification of records by Kerr-McGee which she would leak to the newspaper. During the 30 mile drive to this meeting her car left the road and, at the age of only 28, she was killed. It has been claimed that her death was not accidental, and it is curious that her folder of evidence and notebook against Kerr-McGee have never been found.

In May 1979, after a lengthy legal battle, an Oklahoma jury vindicated Karen Silkwood. They found that Kerr-McGee was 'negligent in its operation of the Cimarron facility so as to allow the escape of the plutonium from the facility and proximately cause the contamination of Karen Silkwood', quoted from Rashke (1982). The jury awarded her estate $505,000 in damages plus $10 million in exemplary damages. Subsequently, she has been called 'America's first nuclear martyr'.

3.5 Rainbow Warrior

There is now irrefutable evidence that governments will resort to terorism to frustrate and prevent legitimate anti-nuclear protests. On 10th July 1985, the DGSE, the French foreign intelligence service, blew up the Greenpeace ship, Rainbow Warrior, while it was in Auckland harbour, killing one person. The Rainbow Warrior was about to sail to disrupt French nuclear weapons tests at Mururoa. Initially the French government denied all responsibility, and an official French inquiry endorsed this position, (Housego, 1985). Shortly after the incident Charles Montan, a French government spokesman, said that 'in no way was France involved. The French Government does not deal with opponents in such ways', Sunday Times Insight Team (1986, page 40). On 22nd September 1986 Laurent Fabius, the French prime minister, admitted that the ship had been sunk by French agents acting on the orders of the French government. Thus the government of a western democratic state engaged in sabotage to stop a peace group from protesting against nuclear weapons testing. This government then engaged in a cover-up and denied they had any part in this illegal action.

3.6 National Radiological Protection Board

The NRPB was established in 1970 by the Radiological Protection Act. It was set up to provide a national point of authoritative reference on radiological protection. The NRPB is authorised to advance knowledge about protection from radiation hazards and to provide advice to those with responsibilities for protecting people from radiation.

The Flowers' Royal Commission on Environmental Pollution (RCEP) called for "expert, open and independent" agencies in radiological protection and for independent assessment of the ICRP recommendations on safety standards. This task has been assigned to the NRPB, yet the NRPB was criticised for its lack of independence even by the diplomatic RCEP, according to Wynne (1978). For example, the NRPB has strong connections with the U.K. Atomic Energy Authority (UKAEA) and Wynne (1978) reports that four of the NRPB's five top scientists were ex-UKAEA scientists, i.e. poachers turned game-keepers.

In 1975, the Guardian newspaper drew public attention to the fact that several workers at Windscale (Sellafield) and other plants where plutonium was extracted or fabricated, had developed leukaemia. The NRPB and

BNFL, denied that any deaths attributable to plutonium had occurred. However, the RCEP felt the facts needed to be clearly established and they asked the NRPB to investigate and produce a public report, RCEP (1976, page 26). In September 1975, the NRPB produced a study of deaths from leukaemia at Windscale (Sellafield) and they concluded that their initial reaction was correct; there was no connection between leukaemia and working at Windscale (Sellafield). The RCEP reported that the NRPB appeared to regard the matter as closed. However, the data used by the NRPB was limited to observations on men only whilst they were employed at Windscale (Sellafield) and so omitted at least half of the deaths from cancer. This introduced a basic epidemiological error into the study since cancer takes many years to develop. It is estimated by the RCEP that by omitting to include cancers in workers after they left Windscale (Sellafield), half of the cancers were missed. The RCEP found it difficult to understand why it was not possible to carry out a proper study of all radiation workers, whether or not they had ceased employment, as has been commonplace in other industries. The RCEP (1976, page 95) go on to state that "we do not think that the NRPB followed up this matter initially with the necessary thoroughness".

Stott et al (1980, pp.139-140) stated that an NRPB statistician, Goss, felt that the findings of the Windscale (Sellafield) study were significant enough to justify further work. This led to a disagreement within the NRPB and Goss resigned. At the Windscale (Sellafield) Inquiry in 1977 the NRPB study was presented as written evidence. It was stated by Stott et al (1980, pp.153-154) that in evidence to the Windscale (Sellafield) Inquiry, Alesbury showed the NRPB report contained methodological and arithmetical errors which led to a significant understatement of the correleation between cancers and radiation exposure in the Windscale (Sellafield) workforce. Dolphin, the author of the NRPB study, admitted the arithmetical errors. In her evidence Dr. Alice Stewart criticised the NRPB report as containing grave defects and inappropriate statistical tests. Some of these errors had been pointed out by Goss in a letter to the Observer newspaper in March 1977, but the NRPB had not withdrawn or corrected their report before submitting it as evidence to the Windscale (Sellafield) Inquiry.

Goss (1977) has written that whilst working for the NRPB he formed the view that "the management were biased towards underestimating radiation risks', were not independent of the UKAEA and that "to make any overt criticism of the management was generally recognised as providing a clear invitation for further intimidation".

During the Windscale (Sellafield) Inquiry it was revealed that the Reverend Postlethwaite, vicar of Seascale for 9 years, had become concerned about the considerable strain on the workforce because of the anxieties over health risks felt by those who did not work at Windscale

(Sellafield). He made known his concern over the resulting low morale of the workforce and safety to the NRPB. The NRPB wrote to BNFL (the operators of Windscale (Sellafield)) stating that they "do not think he (Postlethwaite) will cause you any difficulty". This destroyed the Reverend Postlethwaite's confidence in the NRPB as a source of independent advice, Stott et al (1980, page 123).

The NRPB has been criticised for its partiality on behalf of the UKAEA in the Windscale (Sellafield) Inquiry and for its remarkable zeal in attacking any scientific research which concludes that risks are higher than the ICRP currently maintains. The NRPB's allegedly servile reflection of ICRP attitudes has been criticised. In Wynne's view, the NRPB's assertion that radiation doses below a certain low level, and doses beyond a certain period in the future should be ignored, is an attitude to be expected from the nuclear industry, not from a body which is supposed to be expert, open and independent. It appears that rather than protecting the public from the nuclear industry, the NRPB is protecting the nuclear industry from the public.

3.7 Summary

This chapter has pointed out that rather than being a consensus of informed opinion on the risks from radiation exposure, there is heated controversy. The existence of this controversy is supressed in government documents, where international standards are relied upon as being universally accepted. The dispute within BEIR shows that the controversy has now spread to those organizations which pronounce upon radiation exposure risks. This has occurred in spite of official opposition to research which contradicts the existing estimates, and the possible detrimental effect that conducting such work can have on a researcher's career. Later chapters of this book summarise the work which researchers have managed to conduct and publicise, questioning the existing estimates of the risks of exposure to radiation. Doubts have been expressed about the freedom of nuclear workers to criticise safety procedures and about a lack of enforcement of existing radiation exposure standards. Finally the NRPB, which is meant to be an impartial official body providing advice on the risks of exposure to radiation, was found to devote its energies to defending the status quo against those who wish to see more stringent levels of radiological protection.

4 The difficulties of epidemiological evidence

4.1 Introduction

The figures for the dangers of low level radiation are based upon empirical studies, therefore the validity of such studies is of fundamental importance for this book. This chapter outlines some ways in which the effects of radiation on the cancer rate can be estimated. In particular, the difficulties of conducting such studies will be discussed. This leads to a consideration of the problems in interpreting the results of these studies. A discussion of the difficulties involved in the recognition of occupational diseases follows. These difficulties have impeded the recognition of the full dangers of many toxic agents, including radiation. Finally, some of the epidemiological difficulties of estimating the risks from exposure to plutonium are discussed.

4.2 Epidemiological Difficulties

Controlled experiments where healthy people are deliberately irradiated in order to observe the numbers dying of cancer are unacceptable. This rules out the obvious scientific methodology for determining the dangers of radiation. The results of experiments on animals may not be a good guide to human risks, and the size of any discrepancy is uncertain. The use of a theoretical model of the mechanism by which radiation causes cancer in humans to deduce the risks is not a helpful approach because the relevant mechanism is only poorly understood. So, of necessity, the evidence of risks to humans must come from epidemiological studies of human populations. Epidemiologists do not perform scientific experiments. Instead they use available data on humans who have been exposed to radiation for some other reason e.g. in the course of some medical procedure, and try to draw conclusions about radiation risks from this real world data. Many such epidemiological studies are summarized in later chapters of this book. However, epidemiological studies of radiation risks face a number of difficulties, so making accurate measurement of the risks, particularly at low doses, a difficult task.

A feature of the situation which leads to many of the difficulties, is that cancer can be caused by other factors besides radiation, and there is no way of distinguishing between those cancers caused by radiation and those with other causes. This would not be such a serious problem if the association between the causative factor of interest (radiation) and the effect (cancer) was itself obvious. For example, when someone in a car is instantly killed in a road traffic accident, in the mortuary it is impossible to tell from the injuries in which make of car they were travelling. But it is still possible to discover, with great accuracy, the answer to this question because the effect (death) is instantaneous, and from the wreckage the make of car concerned is obvious. The association between radiation and the subsequent cancer is not apparent for a number of reasons:-

(a) since radiation cannot be detected by human senses, the size of the dose received (if any) may be unknown,

(b) a dose of radiation does not inevitably cause cancer and so the relationship is stochastic i.e. subject to chance,

(c) there is a very long time lag between exposure to radiation and any resulting cancers. This latent period may vary from a few years to over fifty years. This means that the cancers caused by a common exposure of a group of people will be spread out over many decades; Mack et al (1977).

The consequence is that the connection between radiation and cancer can only be established by complicated statistical inference. This means it is impossible to separate people into those whose cancer has been caused by radiation and those whose cancer has some other cause. All that can be done is to estimate the total number of extra cancers caused by radiation. For example, if a linear dose-response curve is fitted the result may be as shown in figure 4.1 and the extra number of cancers caused by a radiation dose of OC is estimated to be the distance AB.

This lack of an obvious connection between cancer and the radiation exposure that was the cause has permitted misleading, although factually correct, statements to be made. For example, in the following statement Mole (1981) compares accidental deaths in the coal mining and nuclear industries. 'The number of occupationally caused accidental deaths in a single year among the employees of the National Coal Board in the UK is

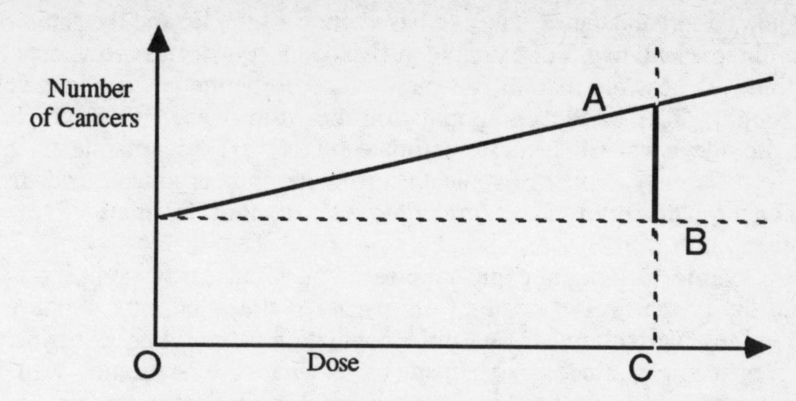

Figure 4.1 Cancers Calculated Using a Dose-Response Curve

greater than all the occupational fatalities from acute radiation effects in the past 35 years in all countries of the world put together.' However, in addition to those who died from acute radiation effects, there have been an unknown, but much larger, number of deaths caused by low levels of radiation. Mole has chosen to omit these deaths from his comparison, presumably because the number cannot be precisely quantified.

Similarly, when discussing the Windscale (Sellafield) fire of 1957 and the Three Mile Accident of 1979, the NRPB(1981) state that 'nobody was acutely injured in either case'. The key word in this statement is 'acutely' since it does not rule out subsequent cancer deaths. Indeed, evidence is presented in chapter 9 to the effect that these two accidents probably did kill people. The UKAEA have gone even further in stating that ' in the worst accident in a commercial nuclear power station, at Three Mile Island in the United States, the built-in protective features ensured that only small doses of radiation were received and no-one was killed or injured',UKAEA(1982). This implies that no cancers were caused by the accident. Not only is there no way of being certain that cancers were not caused, but evidence is presented in chapter 9 that cancers and stillbirths probably were caused. The NRPB(1981) have stated that 'some thirty power reactors have been designed and constructed in Great Britain and have been operated for almost 500 reactor-years without an accident involving serious exposure of staff or members of the public'. The validity of this statement hinges upon what is considered to be a serious accident. Presumably the NRPB does not consider the Windscale (Sellafield) fire in 1957 to be a serious accident, and they take this view because no-one dropped down dead immediately the accident occurred.

There are a number of statistical difficulties in determining whether radiation has caused an increase in cancer rates and, if so, the magnitude of such an effect. A major difficulty is that inferences about the relationship between radiation and cancer must be based upon a sample, and the sample may be unrepresentative. For example, if five death certificates are selected at random it is possible that all five state cancer to be the cause of death. Since roughly 20% of the UK population die from cancer, this sample would provide a very inaccurate estimate of the UK's death rate from cancer. A different sample of five death certificates may produce no deaths from cancer, and again the sample provides a poor estimate of the population value.

It is possible to calculate the precise error when using samples to estimate the proportion of deaths from cancer if the population value is known for any particular year. But if the population value is known the use of samples becomes unnecessary. In situations where the population value is unknown and the sample value must be used as an estimate, the error of the sample estimate is unknown. It is however possible to make an estimate of the likely size of the error. The larger the size of the sample the more representative it is likely to be and the smaller is the likely error. Thus a sample of say five thousand death certificates will probably produce a reasonably accurate estimate of the proportion of the population who die from cancer. However, it has been assumed there are no errors in the cause of death specified on the death certificate. Studies in Japan have found that cancer has been under-diagnosed by 45%, Radford(1983). A further complication is the classification of cancer into its various types, and the fact that the rules for such classifications have been revised over time.

A complication that has not been considered so far is that the rate of cancer deaths varies e.g. from place to place, between age groups and over time, Mack et al (1977). This means that if the sample of death certificates relates to deaths at a particular point in time from a single geographic area for the same age group, the sample value is an estimate of the local cancer rate at that time for the age group concerned. It will, however, provide a less accurate estimate of the overall cancer rate. The greater the variability in cancer rates the more difficult it becomes to estimate accurately the overall cancer rate.

The problem facing epidemiologists is considerably more difficult than just estimating the cancer rate. It is to estimate the *change* in the cancer rate caused by radiation. Excluding the possibility that cigarette smoking causes lung cancers via radiation, which is discussed in chapter 10, only a small proportion of UK cancer deaths are thought to be caused by radiation. So epidemiologists are trying to detect small radiation induced changes in the cancer rate, when there are considerable variations in the cancer rate anyway. These random fluctuations in the cancer rate may be of a similar size to any likely change due to radiation. In consequence it is very difficult to demonstrate a statistically significant relationship between radiation exposure and an increase in the cancer rate.

A further problem is that in many cases the group of people exposed to radiation is not representative of the overall population. For example, the exposed group may be nuclear workers or patients with a particular medical condition, and such groups may well have atypical cancer rates before any radiation exposure. Thus the epidemiologist is not concerned with the change in the overall cancer rate but rather with the change in the cancer rate for the particular group exposed to radiation.

A classical epidemiological technique is the calculation of the standardized mortality ratio (SMR). This compares the actual (or observed) number of deaths (O) with the number of deaths expected (E) on the basis of the death rate from a control population. The SMR is computed as (O/E)x100. The numbers O and E represent sample estimates of the cancer rates of the exposed and unexposed populations. Hence the SMR is the ratio of two estimated cancer rates and so the SMR may have a very high (low) value indicating that the exposed group has a much higher (lower) cancer rate, but this may just be due to sampling error. The likely sampling error of the SMR is reduced by increasing the number of people in both the exposed and the control groups.

The epidemiologist can attempt to answer two questions. First, does the radiation which the exposed group has received lead to a significant increase in the cancer rate? Second, what is the magnitude of any increase in the cancer rate? The first question is logically prior to, and easier to answer than, the second. In the first case the hypothesis being tested is that the change in the cancer rate of the exposed group is zero, whilst the hypothesis being tested in the second case is that the cancer rate has increased by some specified amount. Figure 4.2 shows that for a radiation dose of size OA the estimated increase in the cancer rate is OB. Due to sampling error this estimated increase in risk may be incorrect, but the actual increase probably lies somewhere within limits such as C and D. Since the lower limit in table 4.2 lies above zero, these limits imply that a radiation dose of size OA leads to a statistically significant increase in the cancer rate.

Finding a control group is often a difficult task since the ideal control group does not exist. The ideal is the exposed group in the absence of any exposure and so is hypothetical. The epidemiologist must try to find a control group whose members are similar in all relevant ways to the exposed group so that the cancer rate of the control group is what the cancer rate of the exposed group would have been in the absence of any exposure. However, there will inevitably be some differences between the exposed and the control group because another group with exactly the same mixture of old and young, male and female, smokers and non-smokers etc will not exist. This means it is always open to those wishing to challenge the results of an epidemiological study to argue that the control group used

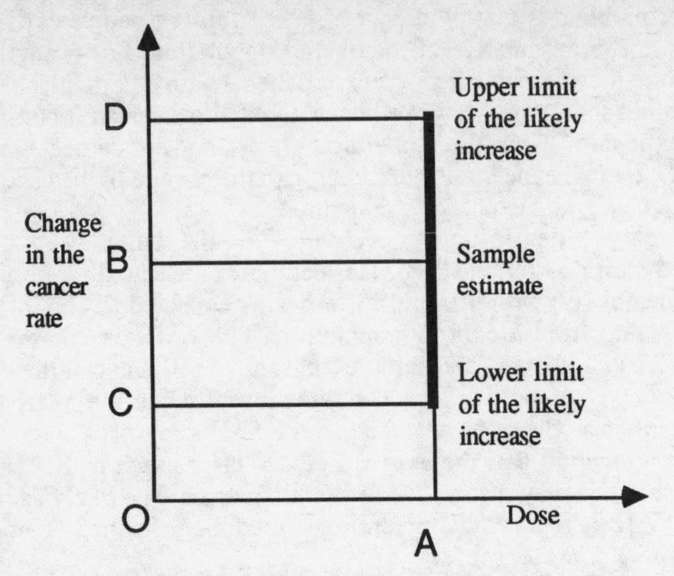

Figure 4.2 Hypothesis Testing and Sampling Error

is not strictly comparable. It is ultimately a matter of judgement whether the control group is sufficiently similar to the exposed group to permit meaningful conclusions to be drawn. For example, various epidemiological studies have been carried out on those suffering from certain diseases who have been given radiotherapy. A suitable control group would involve similar people suffering from the same disease who did not receive radiotherapy. However, most people suffering from a disease will have received the standard treatment, radiotherapy, and so the obvious control group will be too small to produce reliable estimates, Smith(1979).

The additional radiation dose received by each member of the exposed group is usually only known very approximately. Even when the exposed group are wearing badges the internal dose is not measured at all, whilst the external dose is measured only with considerable inaccuracy e.g. 90% errors, GMBATU(1983). If there is no information on the size of the dose, the SMR approach can be used to test whether the unknown exposure has caused an increase in the cancer rate. However, since it is now generally accepted that any additional radiation leads to an increased cancer risk, such a test is of secondary interest.

The purpose of an epidemiological study often is to investigate the effect on the cancer rate of some particular level of radiation. Ideally everyone in the exposed group will have received the same dose, but this is unlikely to be the case. Therefore, what is thought to be the average dose may be used to represent the dose level whose effect is being investigated, but even this limited dose information may be inaccurate. For example, in occupational studies some workers who have not received any extra radiation may be classified as being in the exposed group. This is because in the course of their work they could have been exposed, but in fact were not. This will bias the results towards not finding any significant radiation effect, Mack et al (1977).

There is an additional and more subtle way in which variations in the dose received by the exposed group can lead to an understatement of the risks. If the dose-response curve is concave then, even if the true average dose is known, the greater is the dispersion of the doses the larger will be the understatement of the risk from the average dose. This is illustrated in figure 4.3. If two people both receive a dose of OB the expected increase in the cancer rate is OF. But if one person receives a dose of OA whilst the other receives a dose of OC (with an average dose of OB) the expected increase in the cancer rate is only OE i.e. EF *less* than the increase if they both receive the same dose. Whilst virtually all epidemiological studies of radiation induced cancer involve some variation in the dose, this possible source of bias seems to have been ignored e.g. Gofman(1983, page 88).

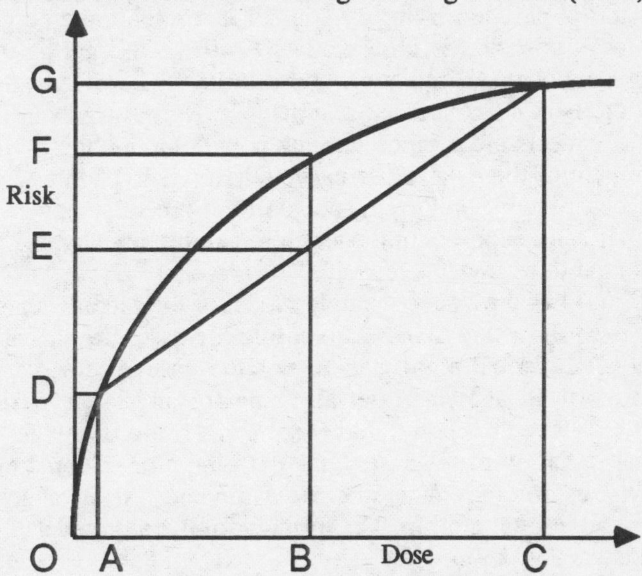

Figure 4.3 Risk Understatement with a Concave Dose-Response Curve

There are two different types of epidemiological study, prospective and retrospective studies. A prospective study is the approach that has been implicitly assumed in this chapter. It involves the identification of a group of people who have been irradiated, and a control group who are similar in all respects, except they have not been irradiated. These two groups are then followed-up and adverse health effects recorded, e.g. the Japanese bomb survivors. The exposed group need not be followed-up from the date of the exposure e.g. the recent studies, described in chapter 7, of troops exposed to nuclear explosions in the 1950's were initiated more than two decades after the exposure.

A retrospective study involves identifying a group of people with a particular disease e.g. leukaemia, and a control group who are identical, except they do not suffer from the specified disease. The investigator then determines whether the disease group has received a higher dose than the control group e.g. the Tri-State study described in chapter 6. The retrospective method is less satisfactory than the prospective method. For example, instead of X-rays causing cancer, people with certain symptoms may be both more likely to be X-rayed and more likely to contract cancer. Hence, even if the cancer group has a higher radiation exposure, the radiation may not necessarily be causing the extra cancers. There is also the problem of reconstructing the radiation doses of each person, and this is often inaccurate and open to bias.

There is a problem with epidemiological studies of the effects of radiation on the population living in a small geographic area e.g. those living close to Windscale (Sellafield). There will be a time lag of many years between exposure to radiation and the diagnosis of any resulting cancers, and the persons concerned may have moved away from the area by the time the cancers are diagnosed. Such population mobility will cause an understatement of the risks, since cancers induced in people who leave the area will be missed. Similarly, if people have recently moved into an area, and so not been exposed long enough to exhibit any adverse effects, the risks will tend to be understated.

Pochin(1976b) has presented a simple model of the factors determining the sample size to show that radiation exposure increases the cancer rate, at the 5% level of statistical significance. For the case where the sample includes people of all ages, and not allowing for any latent period, the required size (x) depends upon the average cancer rate in the absence of any additional radiation exposure (p), the extra number of rems to which the sample has been exposed (r) and the estimated risk of cancer from exposure to an additional rem (k). The formula for the required sample size can now be stated as:- $x = 4p/(kr)^2$.

This formula enables the way in which the various factors affect the sample size to be studied. If the study is concerned with investigating the effects of radiation exposure upon some rare form of cancer (e.g. leukaemia) p becomes small leading to a corresponding reduction in the sample size required. This partly explains why studies have been concerned with the effects of radiation on leukaemia. The formula also shows that if the radiation dose is reduced the required sample size rises. Hence, studying situations where the sample has been exposed to a large dose is attractive, but getting statistically significant results when the radiation dose has been small is difficult. In a similar manner it can be seen that if the estimated risk of an additional rem of radiation inducing cancer (or some specific form of cancer) rises the required sample size drops. This leads to a circularity problem. The epidemiologist is required to know the additional risk of cancer caused by radiation in order to calculate the sample size necessary to show that such a dose does indeed cause an increase in cancer rates. This means that those who think risk factors are high also think that the sample sizes necessary to demonstrate that radiation causes cancer are low. In consequence there can be a difference of opinion over the meaningfulness of some epidemiological study. Making various explicit assumptions about risk factors etc. Goss(1975) has calculated the size of the exposed population required to detect an increase in risk at the 5% level of statistical significance. These are set out in table 4.1. A much larger exposed population would be required to determine the size of any additional risk (i.e. k).

The results of Goss indicate that very large samples are required to study the risks of low level radiation. On this matter BEIR(1980, page 140) have said that "in studies of animal or human populations, the shape of a dose-response relationship at low doses may be practically impossible to ascertain statistically. This is because the sample sizes required to estimate or test a small absolute cancer excess are extremely large". One possible approach to the problem of sample size is to pool data. For example, data on the occupational exposure of nuclear workers in many countries might be treated as one giant sample. But pooling data raises the problem of homogeneity. The doses may differ, the type of radiation may not be the same, the cancer rates in the unexposed groups may differ, etc. Thus, whilst for some types of exposure pooling may be advantageous, each case must be carefully considered.

The relationship between sample size and statistical significance can be used to ensure that significant results are not obtained. For example, an empirical study may group the data by single years of age. In consequence, each year of age would contain only a few observations, and any excess cancers would probably not be statistically significant. Since there would be no single group with a significant cancer effect, it could then be declared that there was no evidence that radiation causes cancer, Gofman(1983, page 162).

The SMR technique requires a control group, and this may be the cause of considerable criticism of a study. So in recent years a statistical technique has been developed which does not require a control group. Based upon the proportional hazards model, it has been called the internal comparison by exposure (ICE) method, Gilbert(1983). It involves comparing the doses received by those in the exposed group who die from cancer with the doses received by a comparable group of exposed people who are still alive. In this context comparable means similar with respect to

Table 4.1 Sample Sizes Required

	Assumed Risk per million persons		Exposed population required	
	Adults	Children	Adults £	Children *
Leukaemia	560	500	100,000	62,000
Cancers other than leukaemia	1,000	500	600,000	62,000
Thyroid	-	125	-	140,000
Lung	600	-	720,000	-
Breast	280	-	2,100,000	-

£ Dose of 20 rads (200 mGy) and an observation period of 20 years
* Dose of 5 rads (50 mGy) and an observation period of 10 years.

age, sex etc. Whilst ICE has the advantage of not requiring a separate control group, it has the disadvantage of requiring accurate data on the doses received, and this is seldom available.

It is often argued by the nuclear industry that just because some epidemiological study finds a positive correlation between changes in the cancer rate and the radiation dose this does not conclusively prove that increases in radiation cause an increase in the cancer rate. In a strict sense this is quite true. Indeed, it is impossible ever to prove *any* scientific theory conclusively, Blalock(1961, ch 1). Bithell et al (1975) state that 'it is logically impossible to conclude for certain from epidemiological observations that a particular association is causal in a specified direction'. Thus the criticism that causality has not been demonstrated beyond all doubt could be levelled against the theories that heat causes water to evaporate or that the moon causes tides on Earth. There is no such thing as a definitive scientific proof and, as Hegel(1807) has argued, all knowledge is subjective. There is always some room for doubt and, if someone is predisposed not to accept a study, reasons for questioning the results can be found. For example, the major tobacco companies still do not accept the overwhelming epidemiological evidence that cigarette smoking causes lung cancer, Royal College of Physicians(1983, chapter 10). Presumably this is because they have a strong predisposition not to believe these results.

Since scientific theories can never be proved conclusively, they are accepted with greater or lesser degrees of confidence in their validity. The scientific method is to test the theory which is initially thought to be correct (the null hypothesis) against the new theory (the alternative hypothesis). The alternative hypothesis is accepted at the 5% level of statistical significance if there is less than a one in twenty chance of the null hypothesis being correct, and the data only supports the alternative hypothesis because of sampling error. It is usual practice for the null hypothesis to be that the variable under consideration has no effect (e.g. that additional radiation is harmless), whilst the alternative hypothesis is often that the variable does have an effect (e.g. that radiation is harmful). This traditional statistical methodology means that harmful agents, such as radiation, are deemed harmless until proven dangerous i.e. they are given the benefit of any doubt. For harmful substances, like radiation, where there are substantial difficulties in collecting conclusive evidence, this benefit of the doubt means that the dangers of some harmful agents will not be accepted without considerable delay and damage to people's health.

If the burden of proof were reversed, so that substances had to prove they were not harmful, this would provide a very powerful stimulus to the collection of much better data. At present those organizations that use radiation have, if anything, an incentive to suppress any data on the health effects of radiation. This is because such data can do them nothing but damage. A null hypothesis that radiation is harmful would be insufficient. It would be necessary to specify the extent to which it causes harm, but there is no obvious number to choose in the absence of any information. No doubt an acceptable solution to this methodological problem could be found; possibly the death rate at which restrictions are imposed on other harmful substances could be used.

The theory that Windscale (Sellafield) has caused a local increase in cancer rates has been criticized on the grounds that correlation does not prove causality. For example, when discussing the high cancer rates in West Cumbria, Black(1984,page 34) states that 'this does not necessarily mean that radioactive waste discharged from the Sellafield site into the atmosphere and sea nearby is the cause of the increase.... An observed association between two factors does not prove a causal relationship. Some third, possibly unthought of factor might be the cause'. But those who use this argument should put forward an alternative explanation which is superior to the theory that Windscale (Sellafield) is the cause. In an annex to chapter three of the report Black considers various other possible causes of the high cancer rates in West Cumbria, and on page 59 concludes that 'we have found no convincing evidence for any unexpected environmental carcinogen or agent around Sellafield'. Thus Black has no alternative explanation to offer (apart from chance), and this leaves radiation from Windscale (Sellafield) as the only reasonable explanation.

A related problem is that of synergistic interactions between agents. For example, each of the constituents of gunpowder is fairly harmless on its own, but when combined in the right proportions the result can be an explosion. Similarly, it is possible that some carcinogens cause only small amounts of cancer when administered alone, but when given in combination the resulting increase in the cancer rate is much larger than the sum of their individual effects. In this context cigarette smoking has been suggested as having an interaction with various other carcinogens in the causation of lung cancer, Smith(1979). The testing of such interactions is difficult since data must be collected on the doses of each carcinogen concerned. It also requires the mathematical form of the interaction to be specified e.g. multiplicative. The limited evidence that is available supports the hypothesis that there is no interaction between cigarette smoking and radiation.

Some types of cancer can have latent periods of over thirty years so that an exposed group must be followed up for more than thirty years if all the radiation induced cancers are to be measured. However, the length of time for which most exposed groups have been followed is considerably shorter than this. Studies which have an inadequate follow-up period will understate the lifetime risks from radiation, BEIR(1980, page 30). For example, some studies have been concerned with the cancers induced amongst radiation workers. Within thirty years of exposure many of these workers change employment or retire, so making it very difficult to follow them up to find out whether they subsequently developed cancer. Similarly, a study of the population in some geographic area must contend with people both moving out of and into the area. The use of cancer deaths rather than incidence further tends to understate the risks because some people, who have been diagnosed as having cancer, will not have died by the end of the follow-up period.

Where a person is subject to prolonged exposure, possibly due to radioactive particles becoming lodged within their body, there is a problem over measuring the relevant dose. Because there is a substantial latent period of variable length for cancers, the irradiation period and the follow-up period overlap and the size of the radiation dose which caused the cancer is unknown. Suppose that a person receives a constant dose of radiation for 25 years, and the cumulative dose is represented by the line OA in figure 4.4. If this person is diagnosed as having cancer in year 25, the problem is to calculate the dose of radiation that produced this result. If the length of the latent period were known precisely (say 10 years) this would be an easy task. The relevant dose would be given by the area of the triangle OBC in figure 4.4. But if, in fact, the latent period was 15 years the dose would have been overstated by the shaded area, DBCE, in figure 4.4.

It is apparent from the above problems that conducting a convincing epidemiological study of the risks from low level radiation is a very difficult task. It is made all the more difficult because much of the data on occupational exposure can only be collected by the nuclear industry itself. Previously this industry has not shown much enthusiasm to collect such data and, when collected, has been very reluctant to release these data to

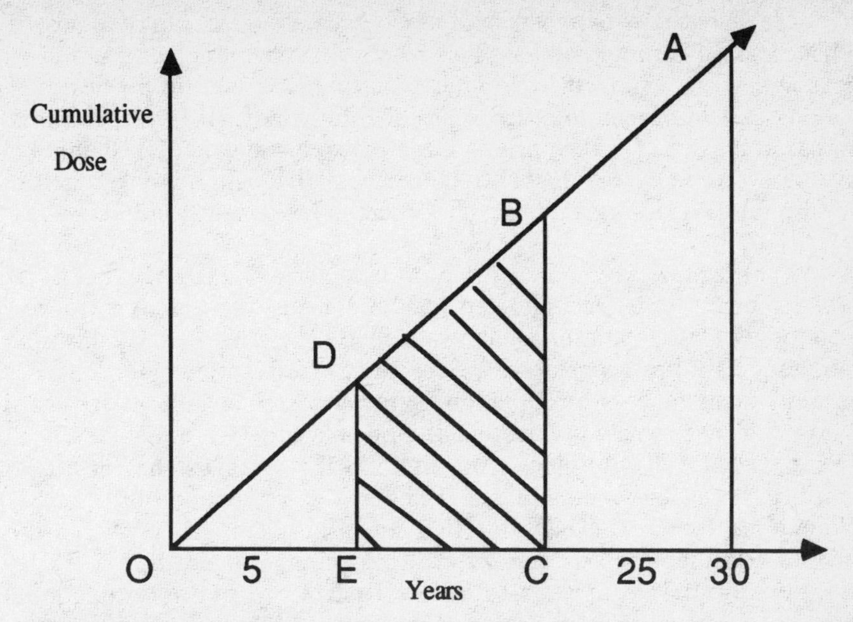

Figure 4.4 Estimation of the Relevant Dose when there is a Continuing Dose

outside researchers, see chapter 3. Failure to collect the data permits the industry to make statements to the effect that there is no evidence to show that there are any appreciable risks from their activities. Of course there is no evidence if no one collects it!

Even when access to the data is not controlled by governmental agencies, the epidemiologist still faces considerable problems. To increase the statistical significance of their results, epidemiologists wish to have a large sample and accurate dose information. Considerable resources may be required to collect the data and, as described in chapter 3, money may not be provided for such studies. In consequence the studies may not be undertaken, or may use a small sample, and so will be unlikely to give statistically significant results.

4.3 Difficulties in the Recognition of Occupational Risks

As well as the above problems with epidemiological studies, there are some more general difficulties which have impeded the recognition of the radiation risks to workers. These problems, which are set out by GMBATU (1983), have previously made the recognition of other occupational diseases more difficult e.g. those caused by asbestos, lead, vinyl chloride, silica and mule spinning oil:-

(a) substances or agents used at work are assumed safe until proven to have caused harm to humans. This means that the burden of proof is on those who wish to show there is a health hazard, and requires the health of workers to be damaged in order to provide this proof. However, workers, not harmful agents (and the profits made by exposing workers), should be given the benefit of any doubt so that preventative action is taken before waiting for a significant number of workers to have their health damaged, Bertell(1979).

(b) it may take some time for sufficient evidence of damage to workers' health to be accumulated for the dangers to be recognized, and even longer for the harmful agent responsible to be identified. By this time there may have been considerable exposure to the harmful agent. This has certainly been the case for radiation. In the sixteenth century it was noted that pitchblende miners in Saxony and Bohemia suffered from a 'mountain sickness'. A local physician wrote in 1556 that 'the dust has corrosive qualities, it eats away the lungs and implants consumption in the body... Women are found who have married seven husbands, all of whom this terrible consumption has carried off to a premature death' quoted by Morgan(1984). This appears to be the first record of an occupational disease associated with pitchblende mining. It took almost another 400 years before the harmful agents were identified, by which time many generations of miners had been exposed. It was subsequently discovered that pitchblende contains uranium and radium, both of which emit alpha particles. Morgan(1984) describes how, after their discovery, uranium and radium were initially used therapeutically. However, by about the 1940's these early therapeutic uses had largely ceased as the dangers were slowly recognized.

(c) even when the evidence is available, the authorities may be unwilling to recognize that a substance is dangerous. This is illustrated by the fact that the average lag in the US between human evidence appearing that a substance causes cancer, and federal regulation, is about 16 years. In 22 of the 26 cases the federal regulation was initiated by petitions, lawsuits, etc. by outside agencies such as trade unions. In the view of Wolfe(1977)

> 'industry and its friends in federal regulatory agencies have often bent over backwards to ignore suggestive or positive evidence of carcinogenicity, as in the cases of many food dyes (e.g. Red 2, Red 4, Citrus Red 2) and pesticides (e.g. endrin, lindane and others). Thus, faced with the regulatory implications of positive scientific findings of carcinogenicity, industry scientists and agency data interpreters have failed to interpret the studies accurately, especially when the findings are positive. This unhealthy fusion of science and regulation, in which studies are interpreted on political and economic grounds rather than solely on scientific or public health grounds, is

difficult for anyone but industry or federal regulatory agency scientists to understand - or even believe'.

(d) the doses that cause the initial recognition that a particular agent is a health risk are often high. For example, early radiologists would hold their hands in front of the X-ray tube to test whether it was working properly, and instead of recording the image on a photographic plate some used fluorescent screens which were examined whilst the patient continued being irradiated. By the time the dangers have been recognized the dose to which workers are exposed has usually been reduced. Since the available evidence relates only to the dangers of high doses, this may be used to argue that the risk from these lower doses is zero or inconsequential. A new campaign must be launched to assess the risks of these lower doses, and an epidemiological study to do this may be considerably more difficult than was the case for the high doses.

(e) the initial high doses may cause fairly obvious symptoms within a reasonably short time of the exposure. These effects are then taken as the *only* health effects. This tends to prevent the recognition of other health effects which have a longer latent period. Thus large doses of radiation produce symptoms indicative of overwhelming damage to the brain and nervous system, or heart failure. Death occurs within about a day. Large doses of radiation e.g.1000 to 3000 rems (10,000 - 30,000 mSv) cause vomiting, diarrhoea, intestinal bleeding, shock and death within about ten days. A dose of about 350 rems (3,500 mSv) will kill half those exposed within thirty days, Pochin(1983, page 102). Fairly large doses of radiation to the skin will cause it to redden and to become inflamed. Hence the effects of high levels of exposure to radiation are fairly obvious and occur within a short period of the exposure. This makes it more difficult to spot the cancers that may be induced up to half a century later.

(f) if the disease has non-occupational causes this can make it more difficult for the occupational risks to be recognized. Cases of the disease amongst workers can be argued to be caused by factors unconnected with their work. This has been a major problem with demonstrating the risks from radiation exposure.

(g) over time the techniques for measuring the level of exposure and for identifying the diseases caused are improved. This leads to the estimated risks being revised upwards because the initial risk estimates are usually too low. This has clearly been the case for radiation where the safety standards have been dramatically reduced over time; see chapter 5.

(h) it may be argued that the early doses were higher than first thought so that the risks are lower than the initial estimates would suggest.

(i) the greater the economic importance of the harmful agent the harder it is to get the risks recognized. Radiation is not only of considerable economic importance e.g. nuclear energy; it is also very important in medicine and for the defence industry i.e. nuclear weapons. The energy and defence policies of a number of highly influential countries rest upon the continued use of nuclear weapons and nuclear energy.

(j) safety standards may be set at levels that the 'dirty' end of the industry can achieve without too much cost. It appears that regulations concerning emissions from nuclear establishments have been set with some regard for the continued operation of existing plants. Thus, as the old plants with high levels of emissions are phased out so the regulations can be tightened.

4.4 Epidemiology and the Dangers of Plutonium

The difficulties with epidemiological studies have permitted various misleading statements to be made. For example, the lack of scientific tests of the health effects of plutonium on humans is sometimes used to make statements to the effect that plutonium has never killed anyone. Spiers et al (1976) wrote that 'no malignancies attributable to plutonium have been recorded in man', whilst Blair et al (1974) have stated that 'no specific physical injury to man has been shown to be caused by plutonium exposure. Many workers exposed to plutonium have been greatly inconvenienced - even pained - by the countermeasures taken to avoid possible effects, and a certain amount of psychological trauma has undoubtedly occurred.' Blair et al appear to be implying that the safety measures taken to protect workers from plutonium have a more adverse effect on the workers than would exposure to plutonium. Subsequently Blair et al make it clear that, if anything, they would like to see stricter plutonium exposure standards. Their statement that plutonium exposure has never been shown to have caused physical injury is most misleading for two reasons.

First, it implies that plutonium is safe, when in fact plutonium is one of the most dangerous substances known to man. This was recognized very soon after plutonium was discovered in February 1941, so that strict safety precautions have been applied to prevent human exposure to plutonium. Because of its lethal nature no scientist has carried out proper tests on humans to quantify precisely the dangers of plutonium. There is ample animal evidence attesting to the extreme dangers of plutonium, although there are difficulties in using animal data to estimate human risks. The only experiments on the human response to plutonium, where the size of the dose is known with certainty, occurred in 1945 and 1946. Eighteen people, who were thought to have a short life expectancy, received injections of tracer doses of plutonium-239. Five of the subjects were lost to the study leaving only thirteen, of which ten are known to have died by 1972, Durbin(1972). So far no cancers of the bone or liver have appeared in any of the thirteen subjects. BEIR(1980, pp 376-377) considered the six patients who survived for over ten years and, using the risk factors for thorotrast (which contains the alpha emitter thorium), calculated that 0.4 cases of liver cancer would be expected. They conclude that this risk estimate is in good agreement with the zero cases observed. This experiment has been used to suggest that plutonium is virtually harmless when, of course, no such conclusion is justified. A sample of six people is far too small to give statistically significant results. In later chapters of this book studies of thousands of people will be presented which have been criticized for having too small a sample.

Second, whilst it is impossible to provide conclusive evidence of a direct causal relationship for the reasons discussed in this chapter, the studies in subsequent chapters of this book give highly persuasive evidence that many people have already been killed by plutonium. The conclusion is that plutonium is known to be a highly dangerous substance that has already killed many people. But for the enforcement of exposure standards from its discovery, ample and irrefutable evidence of its high toxicity to humans would be available.

4.5 Summary

It is logically impossible ever to prove conclusively that a particular substance causes some specified health effect; therefore all such statements are, of necessity, open to attack on the grounds that an absolutely irrefutable case has not been made out. However, it is possible to show that a causal connection is very much more likely to be the case than not. In other areas of science there is a general willingness to accept such probabilistic evidence, but this is not so in the case of radiation. This unwillingness to accept levels of statistical significance which are used in other areas of science is compounded by the difficulties which are inherent in epidemiological studies of the health effects of low doses of radiation. The major statistical problem is to get a large enough sample. This is partly because a large number of people are needed to detect small effects with confidence, and partly because there are difficulties in collecting data on a large sample e.g. the data may not be collected at all, the data may be held by some government body which will not make it available, or there may not be any research grants available to enable researchers to collect and analyse the data. These difficulties have meant that the full recognition of radiation risks has, and is being, delayed. The consequence of this is that misleading statements are made about these risks. The fact that an effect is scientifically hard to prove does not make it any less dangerous. Conversely, all effects that are easy to prove are not necessarily of great importance.

5 Radiation risk estimates and exposure standards

5.1 Introduction

A vital piece of information for those seeking to control radiation exposure is the sources and levels of radiation received by the population. The second section of this chapter sets out the average levels of exposure to radiation for people living in the UK. The institutions involved in setting standards, both internationally and in the UK, are then described and criticised. A key ingredient in setting standards is a set of estimates of the risks involved in exposure to radiation. Section four considers the risk estimates produced by various authoritative bodies, including a discussion of the major revisions to the Japanese bomb data, on which these risk estimates are largely based. Finally, exposure standards and criticisms of them are presented in section five.

5.2 Sources and Levels of Radiation Exposure

The NRPB(1981), has estimated the *average* levels of radiation exposure of residents of the UK from various sources. These are set out in table 5.1.The doses given in the table are averages for the entire UK population. This use of averages can be misleading since some groups of people will receive doses of radiation which are considerably larger than those shown. For example, the doses to local residents from radioactive discharges to the environment by the nuclear power industry during the late 1970's are

Table 5.1 Average Radiation Exposure of the UK Population

Source of Radiation	Rems per Person per Year	mSv per Person per Year
Cosmic *	0.0310	0.31
Geological	0.0380	0.38
Radon Decay	0.0800	0.80
Diet	0.0370	0.37
Total of Natural Radiation	0.1860	1.86
Medical Procedures **	0.0500	0.50
Fallout	0.0010	0.01
Radioactive Discharges	0.0003	0.003
Occupational Exposure	0.0009	0.009
Miscellaneous Sources	0.0008	0.008
Total of Artificial Radiation	0.0530	0.53
Total Radiation Exposure	0.2390	2.39

* Wallace et al (1978) have estimated that a US citizen receives an additional 0.0030 rem (0.03 mSv) per year of cosmic radiation as an airline passenger.
** One chest X-ray gives a dose of about 0.1 rem (1 mSv), NRPB(1981).

estimated by the NRBP (1981) in rems per year to be:-

	Rems per Person per Year		
	Air	Water	Total
Fuel Preparation	0.0005	0.0050	0.0055
Reactor Operation	0.0200	0.0100	0.0300
Fuel Reprocessing	0.0200	0.1300	0.1500

These numbers are very much larger than the 0.0003 rem (0.003 mSv) which is the national average for exposure to radioactive discharges. For some types of occupational exposure the dose is a great deal higher than the average occupational dose of 0.0009 rem (0.009 mSv) given in table 5.1. The NRPB(1981) gives the following estimates in rems per person per year:-

Nuclear Power Workers	0.5500	(5.5 mSv)
Medical Workers	0.2000	(2.0 mSv)
Industrial Radiographers	2.7000	(27.0 mSv)

When discussing average occupational doses, Rotblat(1982) argues that such figures are spuriously low because many workers who have received no radiation dose are included in the average calculation. Similarly Rotblat argues that 'the good image which the nuclear industry is presenting, namely that the radiation hazard is negligible, was arrived at by calculating the dose when spread over the whole world population'. However the radiation dose is not distributed evenly, and some members of the population receive much higher doses than others. The radiation dose from medical procedures is also unevenly distributed throughout the population, with those people who have been subjected to diagnostic or therapeutic procedures using radiation having very much higher doses than the average. The NRPB(1981) appears to have computed the average radiation dose per person by calculating the total radiation dose and then dividing by the number of people. This procedure can lead to some misleading conclusions. It appears to imply that if more people move into an area where there are high radioactive discharges, these discharges become safer because the average dose per person declines!

5.3 Standard Setting Organizations

5.3.1 International Organizations

The premier standard setting organization is the ICRP. In 1925 an International Congress of Radiology was established and had its first meeting in London. The second meeting in Stockholm in 1928 established the ICRP. The ICRP makes recommendations on all aspects of human exposure to radiation. It is independent of any government, and its chairman and twelve members are appointed on the basis of their scientific reputation and standing. They are selected every four years by the ICRP from nominations made by the International Congress of Radiologists and by the ICRP itself. These selections are subject to approval by the International Congress of Radiologists. So the ICRP is a self-appointed body.

In addition to the ICRP, a number of other international bodies make statements on radiation risks.

(a) UNSCEAR was established in 1955 and meets annually and reports on the levels of radiation from different sources and on radiation risks e.g. UNSCEAR(1982).

(b) World Health Organization (WHO) have published standards on drinking water, including the maximum level of radioactivity. This limit was derived from the ICRP recommendations.

(c) Food and Agriculture Organization (FAO).

(d) Nuclear Energy Agency (NEA) of the Organization for Economic Co-operation (OECD). This body proposes radiation standards that are based closely on those of the ICRP. It also organizes the dumping of radioactive waste in the North Atlantic.

(e) The International Atomic Energy Agency (IAEA) was founded as a UN agency in 1957 to promote the peaceful use of atomic energy. It's headquarters in Vienna and it has the power to enforce international treaties and agreements. It also prepares radiation standards based on the ICRP recommendations, in consultation with the WHO, the International Labour Office (ILO) and other bodies. These standards serve as a reference for national legislation.

In addition to these international organizations, the American NAS has a committee on the biological effects of ionizing radiations which periodically publishes reports on radiation risks e.g. BEIR(1980). Their recommendations are advisory and have no legal authority in the UK.

There is one international organization which does have some power to specify radiation standards. European Atomic Energy Community (EURATOM) has the power to fix minimum standards which are binding on the members of the European Community. These standards are based upon the ICRP recommendations and are upper limits. The UK is quite free to set lower limits.

5.3.2 UK Standard Setting Organizations

The setting of radiation standards for the UK is the joint responsibility of the Secretaries of State for the Environment, Wales and Scotland under the Radioactive Substances Act 1960. This is done by means of statutory instruments. They consider the EURATOM standards (which are based on the ICRP recommendations) and, with the advice of the NRPB and the Medical Research Council (MRC), to decide whether the EURATOM or more stringent standards should apply to the UK. Radiation discharges to the environment (atmospheric, liquid or solid) must be jointly authorized by the Secretary of State for the Environment and the Minister of Agriculture, Fisheries and Food. In practice these powers are exercised for atmospheric discharges by the Alkali and Clean Air Inspectorate, and for liquid discharges by the Fisheries Radiological Laboratory.

5.3.3 Criticisms of the ICRP

The setting of radiation standards throughout the world is dominated by the ICRP. It is therefore a matter for concern that the ICRP has been criticized for not being impartial. Parker(1978,page 57) summarized one of the criticisms of the ICRP. This is that the ICRP is 'in effect a largely self elected body, unlikely to elect a person, however well qualified, whose conclusions ran counter to those of its members. It would therefore be opposed to change'. When giving evidence to the Sizewell B enquiry Dr. Rosalie Bertell, a consultant to the US Nuclear Regulatory Commission, accused the ICRP of being an 'old boys network', Guardian (1984a). .

Nearly all of the members of the ICRP since World War II have been employees of governmental atomic energy agencies or worked for governmental radiation related laboratories, Radford(1981a,1981b). This may be partly because the ICRP has only a small budget and so its members must either pay their own expenses for the frequent international meetings or be sponsored by some agency. Thus 'the contribution that the Commission can make to travel expenses is severely limited, and during recent years between one-half and two-thirds of the total travel costs have been borne by the institutions of the ICRP members', ICRP(1977, page 47).

That the ICRP membership is dominated by people who work for nuclear agencies may have led to a lack of impartiality. Thus Radford(1981b) states that 'representatives of governmental Atomic Energy agencies are unlikely to be inclined to reach conclusions that may make development of nuclear energy more expensive or difficult'. In consequence the ICRP might tend to permit greater exposure to radiation than would otherwise be the case. Indeed Radford(1981a) writes of 'the regulators' commitment to the success of the technology whose hazards they were evaluating' and later that ' the ICRP ... now clearly adopt(s) the official positions of the governments for whom most of (its) members work' whilst Morgan(1980), a former member of the ICRP, has written of the 'lack of impartiality by the standard setting bodies and by many who are strongly pronuclear, in evaluating the risks of low level exposure'.

A second major criticism of the ICRP is that it performs two separate functions. It both assesses the risks of exposure and recommends how much exposure is acceptable. The first task is one requiring a detailed and careful analysis of the scientific data, and is best performed by experts. The second task is entirely different since the acceptable level of risk is a social question. For example, Radford(1981b) states that ' it is clear that adoption of any finite risk as acceptable for workers is an ethical, social or political issue, not a scientific one.' whilst Richings(1979) has written that the ICRP 'should have kept strictly to the field where its views are supreme, namely, in the science of relating dose-equivalent limits to particular levels of risk and the techniques of assessing actual exposures. It is entirely a matter for governments to say what risks their citizens should be asked to bear, after whatever consultations they wish to make.'

The ICRP members have no special qualifications to decide what should be society's views. Indeed, because of their background they are unrepresentative of the general population and tend to be willing to accept greater levels of exposure. It is the view of Radford(1981b) that

'there is a subtle source of bias that applies to all scientists who work every day around sources of radiation exposure, and that is that they obviously have concluded that whatever risk they may be running from inevitable exposures, it is worth it to them to have their careers in this interesting and generally well-paid type of work. I maintain that despite their technical knowledge of the subject, this potential bias makes them *less* appropriate as arbiters of what risks *others* should accept from radiation exposure than scientists equally well technically qualified but whose work and livelihood are not so directly related to the day-to-day use of radiation'.

5.3.4 Criticisms of UK Standard Setting

It is apparent from section 5.3.2 of this chapter that the institutional arrangements for standard setting in the UK are complex. Parker(1978,page 59) concluded that the evidence presented to the Windscale (Sellafield) enquiry

'left me in no doubt that a single body would be an improvement on the present situation, for there exists a degree of uncertainity as to where respective responsibilities lie. I would however go further. It would not only be an improvement, it is in my view necessary in the public interest.'

Parker went on to identify a major problem with the existing fragmented system. This is that a person can be exposed to radiation from a number of different sources. Thus the fish they eat and the milk and water they drink may be radioactive, whilst the air they breath may contain radioactive particles and the house they live in may be made of radioactive materials. Parker's view of this situation was ' that an overall view must be taken when the question of the level of discharge to be authorized is considered.' The danger is that separate bodies will independently authorize a set of discharges, each of which is within the standards, but which collectively exceeds the standards.

The NRPB plays an important role in the setting of exposure standards in the UK in that it provides advice to the Secretaries of State on the appropriateness of the EURATOM standards. Unfortunately, the NRPB has been severely criticized for its lack of impartiality, see chapter 3.

5.4 Risk Estimates

A necessary pre-requisite to setting radiation exposure standards is the estimation of the risks. This involves the use of dose-response curves, which have been discussed in chapter 2. Considerable uncertainties are involved in the risk estimates. The statistical difficulties have been considered in chapter 4, whilst the various impediments to conducting empirical studies are set out in chapter 3. The standard setting organizations have based their risk estimates on the Japanese bomb survivors and the ankylosing spondylitis patients. Karl Morgan has criticised this state of affairs. In Morgan(1980) he wrote that

> 'this writer considers it extremely unfortunate that the standard setting bodies, and many who are dedicated to the mission of selling nuclear power, use these two studies as hallmarks of reference and accept them as Gospel truth while going to extreme efforts to depreciate the findings of epidemiological studies that admittedly present some difficult problems of interpretation, but do not suffer from the serious biases that we discussed above'

in section 3 of chapter 2.

5.4.1 The Risk Estimates of Various Reports

A number of organizations have made their own estimates of the health effects of exposure to radiation. Each has had access to the studies that had been published by the time they made their risk estimates, and this may partly explain why their risk estimates differ. Whilst these reports contain a great many figures, they have generally avoided stating some overall risk figure. Since the risks can vary depending on the dose rate, sex and age of the subject, type of radiation, dose, etc this is a sensible procedure. However, when trying to compare the alternative risk estimates, some global measure of risk is very helpful.

To this end various attempts have been made to express the risk estimates of each report in common terms; usually the lifetime risk from exposing one million people with the same age and sex distribution as in Europe and North America to a single dose of one rem (ten mSv). This risk may include or exclude non-fatal cancers and genetic effects. Such calculations may oversimplify the relevant reports but, provided this is recognized, can provide a useful aid to a broad understanding of the risk estimates contained in the underlying reports. For this purpose a summary of the various risk estimates based upon Bertell(1984) is presented in table 5.2. This shows the upper and lower estimates of the number of cancers expected to be induced if one million people are exposed to an additional dose of one rem (ten mSv). The sex and age distribution of these people is that of Europe and North America. Apart from the BEIR estimates which are for a maximum of thirty years, table 5.2 shows the increase in the lifetime risks of cancer.

Table 5.2 Comparision of Risk Estimates

Organization	Lower and Upper Risk Estimates
ICRP(1977)	125
UNSCEAR(1977)	100-200
BEIR(1980)	179-719
Bertell(1984)	549-1,648
Gofman(1981)	3,333-4,255

As well as the overall risk estimates, it is also useful to have some idea how they disaggregate into the different types of cancer. This gives some impression of those cancers that are most likely to be induced by exposure to radiation. Table 5.3 contains the risk estimates given by Bertell(1984, page 15). It is interesting to compare table 5.3 with table 1.3 from chapter 1. Whilst lung cancer is the most common form of fatal cancer, it is cancer of the thyroid that is most likely to be caused by radiation. This type of comparison permits the identification of those cancers that are radiosensitive.

5.4.2 ICRP Risk Estimates

In 1977 the ICRP updated its 1966 recommendations to incorporate the new information that had become available in the intervening decade. The ICRP both estimated the risks from radiation exposure and also recommended exposure standards. Since these are two distinct activities, the ICRP exposure standards will be discussed separately in section 5.5.2. In this section attention will be focused on the risk estimates.

The ICRP risk estimates do not allow for the effects of variations in the dose rate and the age and sex of the subject. As argued in chapter 1, these factors can produce important variations in the biological response, but the ICRP has chosen to ignore them. This is despite the fact that the ICRP themselves have estimated the risk of breast cancer for women as 50 per million women exposed to one rem (ten mSv), whilst for men their estimate is zero. The dose-response curve is assumed by the ICRP to be linear without a threshold, which is described as 'a deliberately cautious assumption', ICRP(1977,page 7). But in chapter 2 of this book it has been argued that linearity may not be cautious assumption. Mathews(1983) has

Table 5.3 Total Lifetime Cancers Expected From a Single One Rem Dose to One Million People

Cancer Site	Expected Number of Cancers
Thyroid	153-242
Liver	97-275
Breast	55-228
Brain	93-186
Pancreas	16-177
Lung	25-150
Skin	10-113
Stomach	33-79
Leukaemia	20-48
Ovary	18-38
Intestine & Rectum	2-44
Pharynx & Larynx	8-19
Sinuses & Mastoids	5-10
Uterus & Cervix	6-8
Bone	2-8
Oesophagus	1-9
Lymphoma	2-6
Kidney	2-6
Salivary Gland	1-2
Total	549-1648

pointed out that the ICRP used the absolute risk model in producing their risk estimates and this gives lower risk estimates than would the relative risk model.

The ICRP estimates of the number of cancer deaths per million people exposed to a whole body dose of one rem (10 mSv) are set out in table 5.4.

Table 5.4 ICRP Risk Estimates

Breast cancer	25
Leukaemia	20
Lung cancer	20
Bone cancer	5
Thyroid cancer	5
Other cancers	50
Total	125

In addition to the above risks of death from cancer, the ICRP estimated that the number of genetic effects in all subsequent generations is 80. However, the ICRP have chosen to confine their attention to the genetic effects in just the first two generations, and estimate this risk at 40. This produces a total risk estimate of 165 cancer deaths and genetic abnormalities when one million people are exposed to one rem (10 mSv). Mathews(1983) argues that the ICRP have completely omitted teratogenic effects e.g. foetal deaths and childhood cancers, even though another ICRP publication in 1977 contains an estimate of these risks. In 1977 the ICRP introduced the use of effective dose equivalents, whereby a dose to an organ is converted to a whole body dose equivalent as described in chapter 1.

5.4.3 BEIR III Risk Estimates

BEIR(1980), known as BEIR III, set out to produce estimates of the excess mortality produced by exposure to small amounts of low LET radiation. They argued there was inadequate data on human exposure to small amounts of radiation and so risk estimates must be based on extrapolation from higher doses. The divisions which took place within the BEIR Committee have been described in chapter 3. Since there is no agreement on the appropriate shape of the dose-response curve for making the extrapolation, BEIR used three alternatives: (a) linear, (b) quadratic and (c) linear-quadratic. Of the dose-response curves considered, the linear model produced the highest risk estimates and the quadratic model the lowest. BEIR also distinguished between absolute and relative risk models. Since BEIR used three alternative dose-response curves and two alternative assumptions about the time profile of the extra risk, they have six alternative sets of risk estimates. This produced a wide range of estimates of the risk resulting from an exposure to radiation. Table 5.5 contains the highest and lowest risk estimates for two types of exposure. It shows the number of deaths from all forms of cancer per million people.

As can be seen from table 5.5, altering the assumptions makes a very considerable difference to the risk estimates. For example, a lifetime exposure to 1 rad (10 mGy) per year, which is quite possible for radiation workers, will kill somewhere between about 50 and 30,000 people, according to which BEIR estimate is used! BEIR(1980, page 354) consider estimates of the increased risk of leukaemia among people exposed to one rad (ten mGy) of gamma radiation and the highest risk estimate is 220 times the lowest risk estimate. In the face of such uncertainty by an authoritative body there is no place for the complacent official view that reasonably accurate estimates of the risks are available.

Table 5.5 BEIR Estimated Excess Mortality per Million Persons

	Quadratic Model Absolute Risk	Linear Model Relative Risk	Difference
Single Exposure to 10 Rads of LET Radiation	95	5,014	53 times
Continuous Lifetime Exposure to 1 Rad per Year of Low LET Radiation	53	28,690	540 times

This wide range of risk estimates has been produced by the BEIR using a linear dose-response curve as the upper limit. However, it has been argued in chapter 2, that in fact dose-response curves may be concave. If the BEIR had used a concave dose-response curve the highest risk estimates, and therefore the range between highest and lowest, would have been even larger.

It should be noted that the BEIR estimates of risk are concerned with low LET radiation e.g. X-rays and gamma rays, and are not directly applicable to alpha radiation, which is high LET. The risks from a similar number of rads of alpha radiation may be twenty times higher, due to its much greater RBE.

Gofman(1983, pp 193-194) has severely criticised some aspects of the BEIR risk estimates. He reports that the BEIR decision that ' for ages under 10 years at exposure, the relative risk ratios obtained appear unreliable, and the ratios for ages 20-29 years at exposure are substituted for them'. In Gofman's view

> 'in a lifetime of scientific research, this author (Gofman) has never seen a more shocking mishandling of scientific evidence than is manifest in this statement. In effect, the BEIR Committee is saying, we are unhappy with the large % excess cancer rate per rad in the 0-9 year age group since it does not have the statistical base we would like, so we will find a group, grossly less sensitive according to every item of evidence available, and substitute the value from that group for the high value that makes us uncomfortable'.

Gofman also criticised BEIR for not taking account of the limited follow-up period of the Japanese bomb data, and for not adjusting properly for the effects of age.

5.4.4 UNSCEAR Risk Estimates

Gofman(1983, page 195) has criticised the UNSCEAR risk estimates on several grounds. In analysing the Japanese bomb data UNSCEAR simply ignored the effects of age. They also failed to allow for the fact that the Japanese bomb data was an air dose and needed to be converted to an absorbed dose. Finally, UNSCEAR failed to allow for the limited follow-up period.

5.4.5 The Japanese Bomb Data

The radiation doses from the two Japanese bombs were estimated by Auxier in 1965 and are known as T65D, which are also known as kerma (Kinetic Energy Released in Material) values. Subsequently, when the basic data was made available, other researchers proposed substantial revisions to these T65D dose estimates. The controversy over this matter is discussed in chapter 3. The T65D doses represent the dose in rads if the tissue were surrounded by air and not shielded by intervening body tissue. The revisions of the radiation doses from the Japanese bombs fall into two main areas; (a) adjustments to the neutron and gamma doses received by unshielded people at different distances from ground zero, i.e. air doses, and (b) alterations in the shielding from radiation of various exposed groups.

The preliminary results of Loewe et al (1981) indicate that at Nagasaki both the neutron and gamma air doses were lower than the T65D figures, whilst for Hiroshima the neutron dose was very much lower and the gamma dose somewhat higher. The neutron dose at Hiroshima at a distance of 1180 meters from the epicentre of the blast is thought to have been overstated by some six to ten times, Marshall(1981a). These revisions result in both Hiroshima and Nagasaki having similar gamma doses and low neutron doses. Previously there was thought to have been a significant neutron dose at Hiroshima. This created difficulties in drawing conclusions from this data, since assumptions about the RBE of neutrons were crucial in estimating the relative contributions of the neutron and gamma radiation. As the Hiroshima data covers more survivors than does the Nagasaki data, this was a serious difficulty. The implications of these dose revisions are that:-

(a) since the radiation at both Nagasaki and Hiroshima is now thought to have been of a similar kind (largely gamma), it is possible to pool all the Japanese data into a single big data set. This will increase the power of any statistical tests which use the Japanese bomb data since the sample size is increased.

(b) since the effects on human health are unaffected by these dose reductions, the estimates of the dangers of gamma radiation are increased.

(c) very little human data will now be available for estimating the RBE of neutrons. Previously the supposedly substantial neutron doses at Hiroshima were used for this purpose.

As well as revising the air doses downwards, it is also possible that the degree to which the population was shielded by buildings has been underestimated. Jess Marcum has concluded that the buildings at both Nagasaki and Hiroshima absorbed more gamma radiation than was previously thought. As a result, this preliminary study indicates that the gamma doses received by humans in both cities will have to be reduced to about 40% of their previous levels. For example, Michael Bender, a radiobiologist at the Brookhaven National Laboratory, was surprised to discover that when the original doses were calculated, a large group of workers at the Mitsubishi steel factory in Nagasaki were simply assigned the dose they would have received if they had been standing outside. Since the Mitsubishi building was made of steel and concrete and contained some heavy machinery, those workers in the factory at the time of the explosion received considerably smaller doses than those assigned to them, Marshall(1981c). This downward revision of the dose data due to shielding is in addition to the downward revisions due to recalculating the air doses. Hence, the shielding revisions are likely to reinforce the effects of the air dose revisions so that gamma radiation is more dangerous than was previously thought.

Some early estimates suggest that, as a result of a reappraisal of the doses at Hiroshima and Nagasaki, the estimated risks from exposure to gamma radiation may be doubled. However, firm estimates are not yet available. More research needs to be conducted on the Japanese dose data, but what is clear is that the recent criticisms of the T65D dose data have 'irreversibly toppled the status quo', Marshall(1981b). The bedrock for estimating the risks from radiation exposure has been smashed. The previous official pronouncements on such risks e.g. ICRP(1977), UNSCEAR(1977), and BEIR(1980), have been invalidated, and a new set of Japanese dose data is required before such bodies can update their risk estimates. For example, Edward Radford, the chairman of BEIR III, has written that 'it is now clear that most of the cancer risk section of the current version of the BEIR III report is obsolete because of new dosimetric data for the Hiroshima atomic bomb', Radford(1981b).

Not only has the dose data for the Japanese bombs been discredited, but new data on the health effects has caused an upward revision of the estimated number of cancers that will be induced throughout the lifetimes of the bomb survivors. The latest data on these survivors has shown an upturn in cancer rates health effects in recent years, Wakabayashi et al (1983). Table 5.6 shows that there has been a most remarkable increase in excess cancers of all kinds except leukaemia in the years 1975-1978. The cancer rate has more than doubled, and this supports the relative rather than the absolute risk model. Table 5.6 contains the number of excess cancers that will appear in one year if one million people are exposed to one rad (ten mGy).

Table 5.6 Excess Incidence at Nagasaki of All Cancers Except Leukaemia

1959-62	7.97
1963-66	6.10
1967-70	7.24
1971-74	9.26
1975-78	19.65

This new data was unavailable to ICRP(1977), UNSCEAR(1977), BEIR(1980) and UNSCEAR(1982). If it had been available it could only have led to higher risk estimates.

Because the Japanese bomb data cannot be used to make reliable estimates of radiation risks, some researchers have used non-bomb data. Charles et al(1981 and 1983) used the data assembled by UNSCEAR(1977) and their estimates are roughly double the value of 125 used by ICRP(1977).

5.4.6 The Burden of Proof

The presumption has been that radiation is harmless until proved otherwise. This has had serious consequences for research into radiation risks. During the 1960's the nuclear lobby stated there was no evidence to show that small doses of radiation were harmful. What they did not say was that there was also no evidence to show they were safe. Whilst drug companies are required to conduct extensive tests to prove a new drug is safe before it is marketed, the nuclear industry was permitted to subject people to radiation doses without proof of the effects of such doses on human health. This reversal of the burden of proof, that low levels of radiation are presumed safe unless someone can clearly prove them to be dangerous, has had major consequences for research into the dangers of low level radiation.

Research into the safety of a new drug is paid for by the drug company concerned, but the nuclear industry had no incentive to commission research into the dangers of low level radiation because the issue had already been decided in their favour. Indeed, they had an incentive to prevent any research in this area since it might prove that low level radiation was dangerous, see chapter 3. The nuclear industry and the government control access to various items necessary for a research project into the effects of low level radiation on humans i.e. data, money, jobs and publicity. The consequence of this situation is that much research into this subject has been suppressed, with many possible research opportunities not being pursued.

It may be argued that this problem was avoided by handing over the task of investigating the health dangers of radiation to various governmental organizations. But where this happened it was ineffective. For example, in the USA the AEC was given the dual roles of both promoting and controlling nuclear power. This put them in an impossible position, as explained in the following extract from Gofman et al(1979, pp 107-108).

> 'It seemed logical, in 1946, to organize a civilian Commission assigned to explore and exploit the phenomena of atomic energy for the fullest benefit of the citizens. The Atomic Energy Commission was given this as one of its missions. But the staggering potential hazard was also recognized and a second mission, that of proceeding with the fullest consideration of protection of health and safety of the public, was also assigned to the Atomic Energy Commission. In this dual mission lay the historic error. No group of people could be expected to do both things at the same time - promote a technology zealously and hastily - and at the same time proceed slowly and cautiously for maximum protection of public health. Go fast but go slowly! This was in essence the directive given the AEC at its inception'.

The consequence of this conflict was that the AEC promoted nuclear power and neglected their duty to protect the public. Thus Lapp(1979, page 85) has written, 'I suspect that the AEC given its way, would have gone on ignoring public concern about radiation risks and would never have launched any large-scale study of its worker force'.

In the UK a Royal Commission has accepted that the burden of proof of safety should rest with the nuclear industry, RCEP(1976, page 99). They state that 'it appears right in principle that operators should bear the onus of responsibility for justifying their releases of radioactivity and showing that these will be innocuous or acceptable'. But subsequently the RCEP report goes on to conclude that, to prevent any duplication of effort, the government and not the nuclear industry should undertake the primary responsibility for conducting research into the effects of radioactive releases.

As in the US, governmental organizations in the UK have not been vigorous in pursuing research that could rebound on their energy and defence policies. It should be noted that, whilst various international organizations such as ICRP, UNSCEAR and BEIR produce weighty reports on radiation risks, they do not themselves conduct any empirical research. Their conclusions are based upon the results of others, and so the quality of their recommendations is limited by the available research. Chapter 3 outlines some of the problems that have been encountered by those who wish to investigate radiation risks in the UK. These include being sacked by governmental organizations, refused research money and denied access to the data.

5.5 Exposure Standards

Setting exposure standards requires the estimation of the health risks and other costs resulting from radiation exposure, and the measurement of the associated benefits. This book, and much of the literature on radiation standards, is largely concerned with the estimation of the cancer risks. However, consideration of the benefits is essential before standards can be set, and this issue will be discussed further in chapter 11. Despite any explicit consideration of the benefits, many authors have set and challenged exposure standards based on different views about the risks rather than the benefits flowing from the use of radiation.

5.5.1 Radiation Exposure Standards Over Time

Whilst it is now accepted that radiation is dangerous, this has not always been the case. For example, in 1917 a medical expert, C.E. Field, said that 'radium has absolutely no toxic effects, it being accepted as harmoniously by the human system as is sunlight by plants', quoted in Lapp(1979, page 46). Although in 1917 this was not the general view, the dangers of radiation exposure were grossly underestimated. Thus at the time that Field made his statement, radium dial painters were receiving doses of radium sufficient to produce a marked increase in their cancer rate; see chapter 8. In consequence exposure standards have changed dramatically over time. The current maximum exposure level for radiation workers is 700 times lower than that suggested in 1902, whilst the public exposure limits in the US are 300 times lower than in 1952. As late as June 1929, the Radio Times was advertizing corsets with a radioactive lining for 30 shillings each, Atom(1985c). In April 1926, in response to a request for advice on the safety of a patent medicine containing radium, the US Bureau of Standards replied that 'this bureau has never heard of any cases of harmful effects due to drinking water which has been made radioactive', quoted from Barlett et al (1985, page 306). Table 5.7 shows the US radiation standards in rems and millisieverts per person per year, and is based upon Morgan(1975a and 1978).

5.5.2 ICRP Exposure Standards

As outlined in section 5.3, national exposure standards are based upon those of the ICRP. In its 1977 report the ICRP established three main objectives:-

Table 5.7 US Radiation Standards Over Time for Workers and the Public

Year	Radiation Workers	
	rem	mSv
1902	3500	35000
1925	46	460
1934	32	320
1949	15	150
1956	5	50

Year	General Public	
	rem	mSv
1952	1.5	15
1958	0.5	5
1974	0.005	0.05

(a) no practice shall be adopted unless its introduction produces a positive net benefit (to society),

(b) all exposures shall be kept as low as reasonably achievable (ALARA), economic and social factors being taken into account, and

(c) the (effective) dose equivalents to individuals shall not exceed the limits recommended for the appropriate circumstances by the Commission.

The ICRP aims to achieve these objectives primarily by setting effective dose equivalents for individual radiation workers and members of the public. The ICRP recommendations are the dose limits which apply in the UK, and are

Radiation worker	5 rems (50 mSv) per year
Member of the public	0.5 rems (5 mSv) per year

In March 1986 the NRPB announced a lowering of the exposure standard for members of the public to 0.1 rem (1 mSv) per year. These doses are whole body doses of the stated amounts i.e. effective dose equivalents. They do not include medical exposure, which may be substantial, nor do they include background radiation. Medical exposures are not subject to any upper limit. However, the ICRP does give three criteria to be used in this context:-

(a) unnecessary exposures should be avoided,

(b) necessary exposures should be justifiable in terms of benefits that would not otherwise have been received, and

(c) the doses actually administered should be limited to the minimum amount consistent with the medical benefit to the individual patient.

The ICRP have used their permitted whole body doses to derive doses to individual organs which are estimated to produce the same risk as the permitted whole body dose. This implies that for a radiation worker the following are the maximum permissible doses to a single organ in one year :-

Breast	33	rem (330 mSv)
Red Bone Marrow	42	rem (420 mSv)
Lung	42	rem (420 mSv)
Thyroid	167	rem (1670 mSv)
Bone Surfaces	167	rem (1670 mSv)
Gonads	20	rem (200 mSv)
Remainder	17	rem (170 mSv)

The ICRP exposure standards permit a radiation worker to be exposed to 5 rem (50 mSv) per year. Using the BEIR relative risk model, a lifetime exposure to 5 rem (50 mSv) per year will produce an 85% increase in the cancer rate. Gofman et al (1979) made a number of plausible assumptions about the cancer risks from low level radiation and concluded that if the US population were exposed to 0.17 rem (1.7 mSv) per year this would induce an extra 100.000 deaths from cancer per year: a 30% increase in cancer deaths. These calculations have been confirmed by the National Cancer Institute and by Nobel laureate Linus Pauling. Whilst the average level of exposure in the US is below 0.17 rem (1.7 mSv) per year, this figure is only one third of the ICRP exposure standard. Hence the ICRP is tacitly consenting to a substantial increase in cancer deaths if the nuclear industry cares to deliver just one third of the permitted dose. Gofman(1976) has estimated that if the US population were exposed to the permitted level of plutonium-239 (and assuming half the adult population to be smokers), these emissions would cause 235,000 extra lung cancers per year just among US males. Calculations of this sort have led to the criticism that 'safety' standards should not permit such large increases in the cancer rate.

5.5.3 The ALARA Concept

In response to criticisms of the above kind, the authorities have argued that the ALARA concept means that doses just under the exposure standard are only permissible if it is not reasonable to reduce them. The ALARA concept (or As Low As Reasonably Practicable (ALARP) as it is known in the UK) has attracted considerable criticism. According to Radford(1981b) 'the 5 rem per year is quickly adopted in the workplace as the "safe" limit, and design and operating practices are tailored to this exposure limit'. Radford refers to the ALARA concept as 'unwise', 'so vague in the legal sense', 'only a pious hope' and 'quite meaningless'. The practical problems in implementing ALARA have been pointed out by GMBATU(1983).

> 'Workers, their safety representatives and the Health and Safety Executive (HSE) Inspectors have great difficulty in challenging a management decision that the exposure on any particular job is as low as is reasonably achievable. Considerable time and resources, both technical and financial, are needed to argue that it is possible to achieve lower exposures than management claims, and that this reduction is worth it in terms of reduced chances of ill-health (the benefit), despite the price of achieving it (the cost)'.

Once the plant has been built the ALARA concept becomes much less powerful. This is because it may be very expensive subsequently to incorporate safety features into an existing plant. It is a great deal cheaper to initially design a plant that is safer, e.g. build an advanced gas cooled reactor rather than a pressurized water reactor. After a plant has been built the ALARA principle may not support its immediate replacement by a safer plant. Therefore society will remain locked into using such a plant, possibly for forty years. This problem of becoming locked into the existing risk levels also applies to the exposure standards themselves. New plant will be designed to comply with the existing exposure standards. Any subsequent lowering of the exposure standards is then unacceptable because it would require the closing down of many existing nuclear facilities.

The ALARA principle requires that the marginal cost of improving safety be compared with the resulting benefits. Whilst the costs can be measured in money, the benefits will be in terms of a reduction in adverse health effects. A comparison between the costs and benefits requires that a monetary value be placed upon these health effects. This is a notoriously difficult problem, and any such valuation is open to challenge, see chapter 11. Thus, the plant operators can argue that the value given to the health effects should be lower, and that therefore the ALARA concept does not require them to improve safety. Hence, what is reasonable becomes a matter of dispute.

In the US, the Navy appears to have taken a different view of what is reasonable from that taken by the nuclear industry. Preventing the exposure of workers servicing reactors is more difficult for nuclear submarines than for civil reactors. The US Navy launched a campaign to reduce exposures, and this not only reduced exposure levels to below those in civil reactors, but also led to a reduction in the total cost of refitting nuclear submarines, Radford(1981b). This illustrates one of the problems with the ALARA concept, since the US Navy appear to have formed a different view of what is reasonable from that taken by civilian reactor operators.

Recently, interest has increased in a new concept which is stronger than ALARA. This is ALATA (as low as technically achievable) or ANZAP (as near zero as possible). The important feature of this new concept is that it does not involve any comparison of costs and benefits. Whenever it is possible to do so risks are required to be reduced. However, ALATA is not part of the ICRP recommendations, nor does it form part of the exposure standards used in the UK. However, according to Atom(1984e), the ALATA concept is supported by Ireland, Sweden, Norway, Iceland, Denmark, Finland and West Germany.

5.5.4 Some Criticisms of Exposure Standards

There is a danger that doses below the exposure limits will be viewed as safe. But, as stated by ICRP(1977, page 16), 'the dose-equivalent limits should not be regarded as a dividing line between safety and danger'. As explained in chapter 2, the true situation is that all doses are dangerous, with large doses being more dangerous than small doses. The risks associated with doses below the standard are deemed to be acceptable, whilst for doses above the standard the risks are unacceptable. However, the view that doses within the exposure limits are safe still persists. For example, in answer to a parliamentary question in March 1985, Mr. Wyn Roberts M.P. commented upon the discharge of radioactive iodine from Singleton Hospital, Swansea, into Bracelet Bay. He said 'that discharges from Singleton Hospital are within the limits laid down in the current authorization which was issued in 1979. There is *no danger whatsoever* from the discharges of radioactive material in use at that hospital', Atom(1985b). Hence, the British government appears to be arguing that doses within the exposure limits are safe, presumably because such doses are below a threshold level. However, the concept of a threshold dose for radiation exposure was generally rejected some fifteen years ago, see chapter 2.

The exposure standards are oriented towards limiting the maximum additional dose to which any individual can be exposed. The size of the collective dose is not subject to the same degree of control. This means that exposing 1,000 workers to 10 rem (100 mSv) per year would contravene the exposure standard for individuals, whilst exposing 10,000 workers to 1 rem (10 mSv) per year would not. In each case the collective dose is the same, but if it is spread across more people the standard that no worker be exposed to more than 5 rem (50 mSv) per year, is met. Using the common assumption that the dose-response curve is linear, the expected number of cancers is the same in either case. This implies that, unless the collective dose is strictly controlled, tougher limits on individual exposure will just lead to the same total dose being given to a larger number of people, with no reduction in the number of cancers. This phenomenon of dose spreading has already occurred; see chapter 8 for a discussion of 'glowboys'.

Whilst the UK has adopted the ICRP recommendations, a number of countries have set lower exposure standards. In West Germany the exposure standard for members of the public is 0.06 rem (0.6 mSv) per year from nuclear installations i.e. just 12% of the UK standard. In the US the exposure standard for members of the public from all sources is only 0.025 rem (0.25 mSv) per year i.e. only 5% of the UK standard, Stott et al (1980, page 141). This raises the question of why, if the US and West Germany can operate successfully with much lower public exposure standards, the UK cannot also adopt lower standards.

The UK uses the ICRP standards for public exposure as upper limits coupled with setting discharge authorizations for each plant on a case by case basis. These authorizations are set after considering the needs of the plant and the capacity of the environment into which the effluents are to be discharged. In contrast, the USA relies on much tighter public exposure limits than those applicable in the UK. The EPA has set whole body limits on the uranium fuel cycle and on wastes that are only 5% of the ICRP dose limit, whilst the NRC design objectives for light water reactor effluent releases are just 1% of the ICRP dose limit. Thus, the limits on public exposure in the US are some 20 to 100 times tougher than in the UK. The relatively lax standards in the UK are defended on the grounds that most emissions are in practice only a tiny fraction of the exposure standard and so actual exposures would meet much tighter limits e.g. those of the US. However, this is not the case. In recent years liquid discharges from Windscale (Sellafield) have approached the ICRP limit and been some 14 times higher than the levels permitted in the US; see table 5.8 which is taken from Hemming et al (1984, page 19).

Table 5.8 Percentage of the ICRP Dose Limit Estimated to be Received by a Member of the Critical Group from Liquid Discharges by Windscale (Sellafield)

Year	Percentage of the ICRP Limit
1978	17
1979	16
1980	16
1981	69
1982	54

This table illustrates the weakness of the UK approach since, if the exposure standard is high, large exposures will not breach the limit. The operators e.g. BNFL can still state that their emissions are within the UK limits even though both the US limits and the levels intended by the UK authorities have been massively exceeded. The Radioactive Waste Management Advisory Committee (RWMAC) has recommended that the critical group should not receive more than 10% of the ICRP's recommended limit, RWMAC(1984, page 41). Table 5.8 shows that the level of discharges from Windscale (Sellafield) clearly breach this recommendation. The Environment Committee(1986, page 158) stated that 'at the Sellafield plant authorised alpha discharge limits were increased in the mid-1950s and again in 1971 to accommodate the increased level of discharge from the plant'. Thus, instead of the discharge limits constraining emissions, the limits were increased to permit the actual discharges. The Committee(page 73) went on to 'recommend that in the Certificates of Authorisation new numerical discharge limits, radically lower than the current ones, should be set for all nuclear plant in the UK. This should be done without delay'.

A number of authors have concluded that exposure standards should be lowered. For example, Schmitz-Feuerhake et al (1978) think that the West German radiation exposure standard of a maximum thyroid dose of 30 rem (300 mSv) per year for those occupationally exposed is much too lax. In their view the exposure standard should be lowered to only 3 rem (30 mSv) i.e. one tenth of its present level. Morgan(1975b) has concluded that the maximum permissible level of internal radiation should be reduced by at least a factor of 200. Schmitz-Feuerhake et al(1979) think that for low LET radiation the cancer risk estimates of the ICRP are between three and seven and a half times too low. Radford gave evidence to the Windscale (Sellafield) inquiry that the exposure limits for workers should be reduced from 5 rem (50 mSv) per year to 0.25 rem (2.5 mSv) i.e. the current standard is twenty times too high, Parker(1978, page 49). Radford also expressed the view to the inquiry that the maximum permissible concentration in air of plutonium and americium should be reduced by a factor of 200. Many other writers have expressed the view that the exposure standards should be lowered by a substantial amount e.g. by five times or more. There are also inconsistencies between various ICRP exposure standards. Using a number of published risk estimates Thorne et al (1976) have calculated the risk of death and hereditary disease from exposure to the maximum permissible body burden (MPBB) of ingested radium-226, strontium-90 and inhaled plutonium-239. They found the ICRP standards permitted a 200 times greater risk from plutonium than from either radium or strontium. They also calculated that to reduce the risk from the MPBB of plutonium-239 to that of 5 rem (50 mSv) of external radiation would require the MPBB for plutonium to be reduced to one fifth of the ICRP standard.

5.6 Summary

The second section of this chapter showed that in the UK roughly 20% of the average radiation dose is due to man-made radiation. (The corresponding figure for the US is considerably higher). Some individuals receive doses that are much higher than the average because they are exposed to additional man-made radiation. Exposure standards are directed towards limiting exposure to man-made radiation since this shows greater variation and is very much easier to control than exposure to natural radiation. There are a number of international organizations involved in setting exposure standards, but the ICRP is paramount. Exposure limits are largely directed against occupational and public exposure to man-made radiation. Medical exposure, which can be very large, is exempt from the explicit exposure limits. The ICRP has been criticised because it is a self-appointed body consisting very largely of scientists working for their respective national governments. In consequence, it has been accused of being biased in favour of the development of nuclear technology. The ICRP, and other standard-setting bodies, perform the two separate functions of risk estimation and standard setting. Whilst they may have special scientific skills appropriate to the first task, they have no particular qualifications to judge what level of risk is acceptable to society. In the UK, the standard setting function appears to be fragmented among a number of organizations and this can lead to problems.

Standards rely on risk estimates, and it has been argued that considerable uncertainty still remains. For example, the recent revisions to the Japanese bomb data and the upturn in the cancer deaths are likely to result in the risk estimates being increased by a substantial amount. The accurate estimation of the risks has not been assisted by the burden of proof being placed on those who sought to show that radiation is dangerous. Past underestimation of the hazards of low level radiation are partly responsible for the massive reduction in exposure standards between 1902 and 1956 in the US. Ultimately, standard setting requires knowledge, not only of the health risks, but of all the marginal costs and benefits associated with additional exposure. A discussion of cost-benefit analysis is postponed until chapter 11, but the ALARA concept, which incorporates a cost-benefit approach, is open to many practical objections. Finally, the ICRP standards have been criticised by many authors for being too lax.

6 Medical exposure to radiation

6.1 Introduction

Only the effects of the exposure of patients to radiation will be considered in this chapter. The effects of radiation exposure upon the health of medical workers (doctors, dentists, radiologists, etc.) who are occupationally exposed will be considered in chapter 8. Generally the consent of the patients is required before they are exposed to radiation so it is a voluntary form of exposure. However, the patients may lack an adequate knowledge of the risks they are running and, if fully informed, might object to being irradiated. Furthermore, doctors, dentists, etc., tend to presume the patient will consent to be irradiated and so the patient must take positive action to opt out of being exposed to radiation. Radiation is used in medicine for two purposes, (a) diagnosis, e.g. X-rays, contrast medium for X-rays; and (b) therapy, e.g. very high doses of X-rays to kill specific cells. In this chapter evidence on the health effects of using radiation for diagnostic purposes will be considered first, followed by a consideration of the therapeutic use of radiation. Whilst the prime source of medical radiation is X-rays (which are external radiation) various radioactive substances are also injected into the body producing internal radiation.

For occupational exposure and the exposure of the public to radiation there are upper limits on the annual radiation dose. However, there are no exposure limits on the radiation dose that can be delivered to patients during medical procedures, and in some cases very large doses are given.

These doses are usually given as part of some therapeutic procedure and are not used for diagnostic purposes. In view of the adverse health effects which even the low doses of radiation used for diagnosis may produce, consideration should be given to specifying exposure standards for diagnostic procedures. Since the medical use of radiation is designed to be beneficial to the health of the patient it is possible to compute the *net* health effects of some medical exposures to radiation, e.g. mammography.

6.2 X-rays of the Foetus

The Oxford Survey represents an important source of data on the health effects of antenatal exposure to small doses of X-rays. It is an ongoing retrospective study where each child dying from a malignant disease in England and Wales has been matched with a control - a healthy child of the same age and sex resident in the same region of the U.K. The data is based upon interviews with the children's mothers using the same interviewer for each child of a pair. Public health departments throughout the country were contacted and the interviews conducted by either principal or assistant medical officers of health. Records of antenatal irradiation by X-rays were verified as far as possible by reference to the records of antenatal clinics, general practitioners and maternity hospitals, Bithell et al (1975). The initial survey in the mid 1950's was largely financed by the Lady Tata Memorial Trust.

Using the Oxford Survey data, Alice Stewart and others claimed in 1956 that antenatal irradiation doubled the risk of the child developing cancer before the age of 10 years. This claim was 'greeted with outraged incredulity' by the scientific community according to Mole (1975b). The Oxford Survey was subject to close scrutiny and a number of criticisms were made. In particular, the retrospective nature of the Oxford Survey, with its reliance on the mother's memories, was criticised. Therefore, what amounted to a prospective study was undertaken by MacMahon (1962) in the north east of the United States to rebut this criticism. He traced the number of cancers and leukaemias among almost three quarters of a million children born in 37 large maternity hospitals in the years 1947-54, and found a 42% increased risk amongst those children subject to antenatal irradiation. This study is seen as confirming the results of Alice Stewart.

Subsequently, Stewart et al (1970) reanalysed the updated Oxford Survey data with information on 7,649 pairs of children born between 1943 and 1965. This data was analysed with respect to whether the mother had been X-rayed during the pregnancy. They found that antenatal radiography increases the risk of death from cancer during the ages 0 to 9 years. Stewart et al (1970) estimated that if one million children were exposed shortly before birth to only 1 rad (10 mGy) of X-rays there would be an extra 570 deaths from radiation induced cancer before the age of ten years. There is some uncertainty about the dose each child received, since the records only give the number of films taken of each child. According to UNSCEAR the average radiation dose per child was less than 2 rads (20 mGy), Pochin (1976a). Hence the studies of antenatal radiation are concerned with low level doses which are well within the occupational exposure standards.

Holford (1975) has reanalysed the Stewart et al (1970) data using a different statistical technique. He concluded that one rad (10 mGy) of antenatal radiation increases the risk of juvenile cancer by between 50% and 100%. Bithell et al (1975) have further analysed the Oxford Survey data. They used data on 8,513 matched pairs of children, where one of the children in each pair died of cancer between 1953 and 1967. The data used included cancer in children up to the age of 15 years. They controlled for various factors - age at death, sex, birth rank order, maternal age, social class, type of region of residence, maternal illness during the pregnancy and type of cancer. It was found that, whilst there were various degrees of association, these factors were not sufficient to explain the observed risk. They conclude that obstetric radiography increases the risk of childhood cancer by 47% and that standardization for various factors has little effect upon this figure.

Whilst a connection between antenatal irradiation and childhood cancer has been established, it has been argued that this does not prove that antenatal irradiation causes childhood cancer. Clinical considerations lead to the antenatal radiation of some children and not others and this could lead to the irradiation of particular foetuses with an above average risk of developing childhood cancer, quite apart from any exposure to X-rays. In consequence, two lines of argument have been developed which refute the view that those children who were selected for antenatal X-rays were predisposed to developing childhood cancer spontaneously.

Some children will spontaneously develop cancer. It is argued that those children who will develop cancer spontaneously are more likely to die at birth or shortly afterwards if they are subject to a difficult birth. Hence the proportion of children spontaneously developing childhood cancer amongst the 'difficult birth' group will be lower. Since the group of children subject to an antenatal X-ray contained a disproportionate number of 'difficiult births', it is predicted that the X-rayed children will have a low rate of spontaneous childhood cancer. Stewart (1973b) reports that 22.5% of the children who died at birth had been X-rayed whilst only 11% of those who survived had been X-rayed. Landau (1974) states that Stewart found a lower cancer rate amongst the irradiated children during their first five years of life. MacMahon only found an excess of cancers amongst his irradiated group after five years of life. These findings are consistent with the view that the irradiated and non-irradiated children are equally likely to develop cancer spontaneously. They also mean that the higher cancer rate amongst the irradiated group is only due to extra cancers developing in the second five years of life illustrating the importance of a long follow up period.

The second line of argument involves an analysis of twin births. Using the Oxford Survey data Mole (1974) found that 55% of the twins had been X-rayed in utero compared with only 10% of the single births. The twins were X-rayed only to confirm the twinning and not for other medical reasons, as was the case for single births. Over half of the twins were X-rayed so that the irradiated group was not some specially selected small sub -group, as in the case of single births. Mole chose to concentrate on non-identical twins because they "will have the same genetic diversity as single births and their genetically determined predisposition to cancer would be expected to be broadly similar unless there is some influence of twinning qua twins". Mole's data is summarised in table 6.1.

Table 6.1 Antenatal Radiation and Twin and Single Births

	Single Birth	All Twins	Non-identical Twins
Number of live births (in 000's)	14,772	353	127
Still born rate	2.1%	5.7%	4.0%
Relative Risk*			
Leukaemia	1.5	2.2	1.5
Solid cancer	1.5	1.6	1.5

*Risk of childhood cancer in the irradiated group relative to the non-irradiated group.

Mole found that those twins who were irradiated had a much higher cancer risk than those who were not. For non-identical twins the relative risk for both leukaemia and solid cancer is the same as for single births, whilst for all twins it is higher. This indicates that for twins, where the irradiated group are not a specially selected sub-group, the exposure to antenatal radiation causes an increased rate of childhood cancer. BEIR (1980, page 444) conclude that Mole's results indicate a causal connection between antenatal radiation and cancer.

Using the Oxford Survey data Stewart (1973b) found that the childhood death rate from cancer amongst twins is only 69% of that for single births. She concluded this is not because twins are less sensitive to carcinogens, but rather it is because those twins who are cancer-prone are more likely to die at or before birth than are singletons who are cancer-prone.

The work of Alice Stewart has led to the eventual acceptance that X-raying pregnant women causes a significantly increased risk of cancer to the child. This has resulted in a considerable reduction in the X-raying of pregnant women in the abdominal and pelvic region. It appears that foetuses have a much greater sensitivity to radiation induced cancer than children or adults, and that the sensitivity is greatest early in the pregnancy. BEIR (1980, page 448) estimate that the risk of cancer is increased by 400% if the X-ray takes place in the first three months of pregnancy and 47% if the X-ray is later. This great sensitivity in the early months of a pregnancy has led to the development of the 'ten day rule'. In 1966 the ICRP recommended that

> 'the ten-day interval following the onset of menstruation is the time when it is most improbable that such women could be pregnant. Therefore, it is recommended that all radiological examinations of the lower abdomen and pelvis of women of reproductive capacity that are not of importance in connection with the immediate illness of the patient, be limited in time to this period when pregnancy is improbable.'

Some doctors have taken the risks of obstetric radiation very seriously, and have gone very much further than the 'ten day rule'. Neumeister (1978), head of a centre established by the East German government to investigate the effects of radiation on the foetus, has recommended that if the foetus has been exposed to over 10 rads (100 mGy) the pregnancy should be terminated. This is presumably because of the increased risk that the child will be deformed, suffer from cancer and be prone to infection.

The difficulties in conducting epidemiological studies to investigate the risks from low level radiation were discussed in chapter 4. Mole (1975b) has pointed out that Alice Stewart and others were only able to show that ante-natal X-rays cause cancer because of very special circumstances:-

(a) childhood cancer is normally very uncommon,

(b) exposure to antenatal X-rays is memorable and can be recorded,

(c) the whole body of the embryo is irradiated so maximizing the chances of producing cancer, and

(d) a population numbering millions of children could be examined over a period of years.

6.3 Tri-State Leukaemia Survey

The Tri-State Leukaemia Survey, which is described by Graham et al (1963), covered all those who were diagnosed as having leukaemia or who died from leukaemia between 1959 and 1962 in 26 counties in upstate New

York and the counties in and around Baltimore, Maryland and Minneapolis-St.Paul, Minnesota. The population base for the study was 13 million people. Controls were randomly selected from a sample of households within the same geographical areas as the leukaemia cases. The controls were selected to produce the same age distribution as the leukaemia cases. A summary of the Tri-State data is set out in table 6.2.

Table 6.2 Tri-State Leukaeumia Survey

	Male	Female	Total
Leukaemias	850	564	1,414
Controls	668	702	1,370
Population	-	-	13,001,970

Each subject or members of his or her immediate family were interviewed. Intense efforts were made to confirm all the data on past X-rays. Inquiries were made to all doctors, dentists, chiropodists, hospitals and radiology departments which had ever treated the subject or his or her family to produce as accurate a history of irradiation as possible. The Tri-State Leukaemia Survey data has been used in a variety of studies of the effects on health of low level radiation.

6.3.1 Premature Ageing

A study of radiologists reported in chapter 8 has found that occupational exposure to radiation shortened their lives. Bertell (1977) hypothesised that exposure to radiation accelerates the natural ageing process, so that the observed increase in various chronic diseases among those exposed to radiation is normal for their altered biological age. Bertell chose to test this hypothesis using the Tri-State Survey data on nonlymphatic leukaemia. She defined ageing or life shortening as 'the stress on the individual which undermines his biocomunication facility, reducing his ability to cope with life and maintain health homeostasis. Such breakdown is associated with debilitating illness prior to death and actual premature termination of life'.

Bertell posits that 'exposure age' is equal to chronological age plus the cumulative radiation dose (in rads) multiplied by some constant (K). She then tests the hypothesis that people of the same exposure age have the same probabilitiy of contracting leukaemia. For example if K equals one, a person of 30 years of age who had been exposed to 15 rads (150 mGy) would have the same chance of contracting leukaemia as a 45 year old person who had not been exposed to any radiation. Using the data for males aged over 45 years, Bertell found that the value of K which best fitted the data was 0.6 i.e. exposure to 15 rads (150 mGy) increases the 'exposure age' of a 30 year old person to 39 years.

6.3.2 Dose-Response Curves

The Tri-State data has also been used by Bross (1979) and Bross et al (1979). Dose-response curves were constructed for X-rays and the excess risk of leukaemia over the range 0 to 40 rads (400 mGy) for men of different ages was calculated. It was concluded that at low doses the dose-response curve tends to flatten out, ruling out the possibility of any threshold dose below which radiation is harmless. The excess risk of leukaemia in men of different ages is shown in table 6.3.

Table 6.3 Excess Risk of Leukaemia in Men of Different Ages

Dose	Age		
	15-44	45-64	Over 64
Under 1 rad	174%	31%	37%
1-5 rads	174%	125%	50%
5-10 rads	97%	89%	48%
10-20 rads	305%	249%	206%
Over 20 rads	305%	417%	549%

The new statistical methodology employed by Bross et al (1979) has been criticised by Boice et al (1979).

6.3.3 Trunk X-Rays

Gibson et al (1972) used the Tri-State data and analysed 1,414 adult leukaemia cases and 1,370 adult controls. For males they found considerable increases in cancer rates with exposure to X-rays.

Table 6.4 Relative Leukaemia Risk to Males from Trunk X-Rays

Number of X-rays to the trunk	Acute Lymphatic Leukaemia	Acute Meloid Leukaemia	Chronic Myeloid Leukaemia
11+	1.63	1.72	2.22
16+	1.75	1.72	2.33
21+	-	2.23	3.26
41+	-	6.78	7.14

The above table shows the observed numbers divided by the expected numbers. This shows that for 21+ X-rays there is a two to three fold increase in myeloid leukaemia, whilst for over 41 X-rays there is a seven fold increase. Brown (1976) has argued that if a dose of 0.5 rads (5 mGy) per X-ray is assumed, these data yield a risk estimate of roughly 10 cancers per million people per year per rad, which is an order of magnitude higher than estimates obtained from high doses. This implies that the dose-response curve is concave.

6.3.4 Groups Susceptible to Low Level Radiation

Bross et al (1972) have used the Tri-State data to examine the hypothesis that some groups are much more susceptible to low level radiation than others. They analysed 295 children with leukaemia and 813 controls and found that if a child had been X-rayed during pregnancy the risk of developing leukaemia between the ages of 1 and 14 years was increased by 40%. However, if the child had suffered from the allergic diseases of asthma or hives as well as being X-rayed in utero the risk of leukaemia was 740% higher than otherwise. If the child had previously had pneumonia, whooping cough or dysentry, as well as being X-rayed, the risk of childhood leukaemia was increased by 310%. Mole (1974) has criticised Bross et al (1972) for basing their study on the entire group of children in the Tri-State Survey whose mothers were exposed to X-rays during the pregnancy, rather than the 30% of that group whose mothers had abdominal exposure.

In another study Natarajan et al (1973) used the Tri-State data to analyse the effects of pre-conception irradiation of the mother (excluding therapeutic radiation) on childhood leukaemia. They re-analysed the 295 children with leukaemia in the Tri-State study and 813 controls and found that preconception radiation exposure did not have a significant effect upon leukaemia risk. However, if the child had suffered from the alergic diseases of asthma, hives or eczema as well as having a mother who was irradiated prior to conception the risk of childhood cancer was increased by roughly 360%. Similarly, if the child had suffered from pneumonia, dysentry or rheumatic fever and if their mother had been irradiated before conception the risk of childhood cancer was increased by about 440%. The authors interpreted this result as being due to pre-conception irradiation damaging the ova leading to the birth of children more susceptible to leukaemia and to allergic and bateriological diseases.

Bross et al (1974) have considered two alternative interpretations of their studies that radiation of the mother prior to conception or during the pregnancy leads to an increased risk of childhood leukaemia. The alternative hypotheses are:-

(a) Certain diseases are an indicator that the child is susceptible to radiation exposure causing leukaemia, or

(b) These diseases are early symptoms of leukaemia and do not indicate an increased susceptibilitiy to radiation exposure causing leukaemia.

Both hypotheses predict higher leukaemia rates amongst children exposed to radiation who have also previously had certain diseases. However, they differ in their predictions about the risks of leukaemia for children who have not been exposed to radiation but who have had the specified diseases. If hypthesis (a) is correct the risks of childhood leukaemia will not be increased, whilst if hypothsis (b) is correct there will be an increased leukaemia rate. Bross et al (1974) used the Tri-State data to distinguish between these two hypotheses and conclude that hypothesis (a), that there is a group of children who are susceptible to radiation causing leukaemia, is supported.

Bross et al (1977) used a different and novel statistical technique to analyse the Tri-State data considered by Bross et al (1972), and concluded that for children who had been X-rayed whilst in utero and had experienced asthma, pneumonia, dysentry, rheumatic fever, etc., the risks of childhood leukaemia were increased by roughly 4,900%. Hence there is a group of children (about 1%) for whom radiation is especially dangerous. This point tends to go unnoticed when averages across all children are used.

A subsequent and more detailed analysis by Bross et al (1980) has confirmed these earlier results. This later study allowed for the possibility that childhood leukaemia is caused not only by interuterine radiation of the child but also by preconception irradiation of the mother and father (i.e. genetic effects). The results support the view that preconception irradiation of the father, the mother and interuterine irradiation of the foetus all increase the leukaemia rate amongst sensitive children.

The studies of Bross and others have shown the existence of groups of people particularly sensitive to radiation. The study of thyroid cancer by Hempelmann et al (1975) referred to in section 6.6 of this chapter also supports the view that the population contains sub-groups who are especially radiation senstive. They found that women and Jewish Americans were more sensitive, with the risks of Jewish women developing thyroid cancer being seventeen times higher than for the non-irradiated group.

6.3.5 Heart Disease, Leukaemia and X-Rays.

Bross et al (1978) have also used the Tri-State data to study the effects of X-rays on the rate of heart disease and leukaemia in men. It was hypothesised that low level doses of radiation (estimated to be between 0 and 50 rads (500 mGy)) caused genetic damage to the DNA of the cells of the blood forming organs, diminishing the effectiveness of the production of some particular enzyme. This is argued to lead to various disorders including leukaemia and heart diseases. Bross et al (1978) found that the risks of heart disease were tripled by exposure to low level radiation whilst the risk of leukaemia was ten times higher.

6.4 Mammography

X-ray examination of the breast (mammography) is used as part of the mass screening of women for breast cancer. However, the wisdom of such mass X-rays has been seriously questioned. Bailar (1976, 1978) has analysed the Health Insurance Plan (HIP) study of 62,000 women aged 40 to 64 years old in Greater New York. Of the 20,166 women who were screened, Bailar estimated that a maximum of 12 were saved from dying from breast cancer as a result of the mammography, i.e. 6 in 10,000. However, these women were subject to 64,810 mammographic examinations, and X-rays are known to induce breast cancer as well as revealing an existing cancer. The average radiation dose to the skin in the HIP study was 7.7 rads (77 mGy) per mammographic examination whilst the dose to the breast tissue was probably over 4 rads (40 mGy). Since each woman received an average of 3.2 mammographic examinations, the total radiation dose received per woman was about 13 rads (130 mGy). Using the NAS risk factor for the induction of breast cancer by X-rays, the HIP study is estimated to have led to a net *increase* in deaths from breast cancer of about 20. In addition to these 20 extra deaths from breast cancer, the X-rays will also have caused some deaths from leukaemia and lung cancer. The implication is that the HIP study was counterproductive in that overall it caused a net increase of over 20 deaths.

Bailar points out that the radiation dose resulting from a mammographic examination has fallen in recent years. He estimates that if a dose of 2 rads (20 mGy) to the breast is used per mammographic examination the number of deaths from breast cancer caused by the X-rays would have been 16, i.e. that if repeated the HIP study would still lead to a slight net increase in deaths from breast cancer. In 1976, Bailar wrote that

> 'the question of possible radiation hazards of mammography is not new...I am unable to explain why questions about the effects of radiation used in mammography have not been investigated more actively, and any well-supported conclusions documented in the open literature.'

There are other costs and benefits apart from those which prevent or cause death associated with the mass screening of women for breast cancer using mammography, e.g. earlier diagnosis permitting less radical treatment, the monetary costs of mammography, the impact on patients of incorrectly diagnosing breast cancer, the possible delay in correctly diagnosing cancer if the patient is incorrectly informed they do not have breast cancer, etc. After considering these other costs and benefits which do not involve the life or death of the woman concerned, Bailar concludes that no woman under 50 years of age should be screened for breast cancer, and in 1978 the U.S. government adopted this as its official policy. This age distinction is made on the grounds that younger women have denser breast tissues which makes correctly diagnosing cancer more difficult (although with modern equipment this is less of a problem), a lower incidence of breast cancer and a longer future lifetime in which to develop X-ray induced breast cancer. It is possible that ultimately mass screening for breast cancer using mammography will cease for women of all ages because of the danger of cancer from the low doses of radiation involved.

6.5 Excessive Diagnostic X-Ray Exposure

According to Morgan (1975a) medical diagnostic exposure in the U.S.A. is two to ten times larger than in most advanced countries. Gofman (1982, page xiii) has estimated that the equivalent of 0.05 rads (0.5 mGy) per year of whole body radiation received by the average American from medical and dental X-rays will cause 46,600 fatal cancers each year. McClenahan (1969) has given numerous examples of situations when unnecessary X-rays are taken in the U.S., whilst Morgan (1973) presents thirty reasons why the U.S. has such a relatively high exposure to medical X-rays and these may be summarized as follows:-

(a) X-rays are used as a routine diagnositc procedure because of the heavy legal penalties for failing to do so and because insurance covers most of the costs of X-ray examinations.

(b) X-rays are required for certain jobs and in law suits for injury.

(c) X-rays increase the income of the doctor or medical institution concerned.

(d) Fresh X-rays are taken instead of consulting previous X-rays.

(e) The X-ray equipment in use is poor and the techniques of operation of these X-ray machines are inadequate.

It was Morgan's view in 1975 that with better training, more modern X-ray equipment and the use of better diagnostic techniques, the average medical exposure in the U.S. could easily be reduced to 10% of its current level. What is more, he argued that this much smaller exposure to low level radiation would produce better X-rays and more meaningful diagnostic information.

Table 6.5, which is taken from Morgan (1975a), summarises the amount of exposure to low level radiation from different types of diagnostic X-ray exposure. This shows that, for some types of X-ray, a number of people will receive a dose that is 100 times larger than the dose received by others. This indicates that very substantial reductions in dose are possible. Further evidence for this is provided from Canada. Taylor et al (1979) measured the skin doses for various X-ray procedures from 30

X-ray machines in hospitals in Toronto, Canada. They also found considerable variation in dose, and these are summarized in table 6.6. By making a number of simple and costless changes in the way the machines were used, Taylor et al were able to reduce doses by between 50% and 70%, with no reduction in the quality of the X-rays.

Table 6.5 Diagnostic X-ray Exposures in the U.S. Measured in Rems to the Skin

	Range of Doses		Average	Upper/Lower
	Lower	Upper	Dose	Dose
Chest X-ray at ORNL (radiographic)	0.010	0.020	0.015	2 times
Chest X-ray in U.S. (radiographic)	0.010	0.300	0.045	30 times
Chest X-ray in U.S. (photofluorographic)	0.200	2.000	0.504	10 times
Dental X-ray series in U.S.	1.000	100.000	20.000	100 times
Abdomen X-ray by a radiologist (radiographic)	-	-	0.636	-
Abdomen X-ray *not* by a radiologist radiographic)	-	-	1.253	-

Table 6.6 Variations in X-Ray Dose Between Machines

Diagnostic Procedure	Upper/Lower Dose
Chest	6 times
Barium Meal	22 times
Barium Enema	5 times
Intravenous Pyelogram	29 times
Gall Bladder	10 times

Morgan (1973) pointed out that, not only did the opportunity exist for a massive reduction in medical exposure in the U.S.A., but that such an opportunity was also present in the U.K. In 1966 it was estimated that if all radiological departments in the U.K. employed the techniques used in the best 25% of the departments in 1958, the population dose from diagnostic radiography would fall by a factor of seven.

The mounting evidence on the dangers of low level radiation has resulted in changes in policy in the U.S. and U.K. on radiation exposure. Morgan (1975a) reports that the mass X-raying of school children in the U.S. was stopped in 1972. In the U.K. the use of pedoscopes (X-ray machines to test the fit of shoes) in shoe shops has ceased. In section 6.2 the increased caution in X-raying pregnant women in the abdominal region was explained, whilst in section 6.3 the U.S. policy of not subjecting women under 50 to breast screening was noted.

6.6 Thyroid Cancer and X-Ray Treatment of Children

Between 1940 and 1959 the New York University Hospital Skin and Cancer Unit treated 4,354 children for tinea capitis (ringworm of the scalp). Until 1959 a widely used treatment was the X-raying of the child's head to cause all the hair to fall out permitting treatment of the ringworm. This involved a dose of radiation ranging from 450 to 800 rads (4500 to 8000 mGy) to the scalp down to only 6 rads (60 mGy) to the thyroid. In 1959 this X-ray treatment was replaced by the use of an antifungal agent. Of these 4,354 children, 2,545 were irradiated whilst 1,809 were treated without the use of X-rays. Shore et al (1976) carried out a follow-up study of these children giving an average post-treatment time of 20 years. Shore et al obtained information on 1,982 irradiated children and 1,258 non-irradiated children. They found a level of tumours of the head and neck amongst the irradiated group that was more than double that of the control group. They also found that the level of psychiatric disorders amongst white children who were irradiated was 40% higher than for white children who were not irradiated. Finally they found that the rate of skin cancer of the head and neck amongst the white irradiated group was over fourteen times higher than for the white control group.

In a study by Modan et al (1974, 197a) 10,902 Israeli children subject to X-rays in the treatment of ringworm of the scalp were retrospectively followed up for twelve to twenty three years. In addition, two control groups of 10,902 non-irradiated children and 5,496 siblings of the irradiated children were also followed up. The average dose to the thyroid was 6.5 rads (65 mGy), and this dose estimate has subsequently been confirmed by later experiments by Modan et al (1977b). Pochin (1976a) estimated that the results of Modan et al (1974) imply that if one million children were X-rayed to assist in curing ringworm 910 would develop thyroid cancer. This amounts to 140 thyroid cancers per rad. It is important to note that the Modan et al (1974) study did not consider any other harmful effects to other parts of the body.

The average dose involved in the study of the Israeli children was 6.5 rads (65 mGy) to the thyroid. However, Silverman et al (1975) point out that there are other medical procedures which involve much higher doses of radiation to the thyroid gland. For example, a thyroid scan using iodine 131 will involve an estimated dose to an adult thyroid of about 150 rad (1500 mGy). Silverman et al (1975) comments that the effects of the use of such diagnostic procedures on children have not been studied systematically. However, Pilch et al (1973) present the case history of a 20 year old girl who was found to have thyroid cancer. At 4 and 12 years of age she was suspected of hyperthyroidism and given radioactive iodine as part of the tests. Since iodine tends to concentrate in the thyroid gland, Pilch et al conclude that the previous diagnostic procedures to which the girl was subject could well have caused her subsequent cancer. They recommend caution in the administration of tracer doses of radioactive iodine to children for diagnostic purposes.

Hempelmann et al (1975) carried out a follow-up survey of 2,872 people who were treated in the U.S.A. with X-rays in infancy. An average dose of 119 rads (1190 mGy) to the thyroid gland was given to shrink an allegedly enlarged thyroid gland. This procedure was thought to alleviate respiratory distress or to prevent the sudden death of a previously healthy infant. As a comparison 5,005 non-irradiated siblings of the treated group were also studied. It was found that over the twenty years following the treatment there were 46 cancers amongst the irradiated group as compared with 12.3 cancers which would have been expected (based upon the average cancer rate in upstate New York), ie. a 174% increase in the cancer rate. For the siblings there were 25 cancers as against 23.6 expected.

The increase in thyroid cancer rates was much more marked. Amongst the irradiated group there were 24 thyroid cancers whilst only 0.3 were expected. This represents an increase of almost 80 times. For the untreated siblings there were no thyroid cancers as against 0.6 expected. This study indicates that in about the year 1950 thousands of infants were subject to high doses of X-rays for theraputic reasons. Hempelmann et al (1975) do not mention whether the treatment had any beneficial effects, but they do show that it led to roughly 34 extra cancers, mainly cancers of the thyroid.

6.7 Cranial X-Rays and IQ

Children who have acute lymphocytic leukaemia (ALL) are usually treated by drugs and cranial X-rays of 2,400 rads (24,000 mGy). Meadows et al (1981) conducted a study at the Children's Hospital in Philadelphia to determine whether this treatment led to a decline in the children's IQ's. They conducted a follow-up study of 18 children with a median age of 4 years 10 months at diagnosis who were diagnosed as having ALL. These children were given three IQ tests, and their median scores are summarised in table 6.7 where a score of 100 represents the average score for all children of the appropriate age. The IQ scores of the 18 ALL children on the first two tests, although having a median of above 100, were not significantly different from 100. However, on the third IQ test, when the median score had dropped to only 89, the scores of the 18 ALL children were significantly different from 100.

Table 6.7 IQ Tests of the ALL Children

Test	Months after Diagnosis	Median IQ Score
1	1	109
2	12 to 34	106
3	32 to 68	89

The drop in IQ was analysed by age and it was found that children under 5 years old, when diagnosed, showed the greatest decline in IQ. All of the children studied attended school regularly after diagnosis and so their impaired performance in IQ tests cannot be attributed to interrupted schooling. In addition, the ALL children studied did not show signs of psychological problems that were sufficiently serious to produce the observed decline in IQs.

Whilst the sample size is small, the authors concluded that the treatment for ALL caused a reduction in the IQ of the children concerned. However, this decline in IQs may be due to the X-ray treatment, the drug therapy, the combination of both or the ALL. To try to decide between these alternatives, the authors tested the IQ of a control group of 6 children who had been diagnosed as having ALL between 7 and 13 years previously. These children had received only drug therapy and had a median IQ of 109. The authors conclude that the cranial X-ray treatment was largely responsible for the decline in the IQs of the 18 ALL children. Although the doses of radiation were substantial, this study indicates that theraputic doses of X-rays can impair the intelligence of children.

6.8 Thorotrast Patients

Between 1928 and 1955, when its use was stopped, thorotrast was injected into patients as an X-ray contrast medium primarily for the diagnosis of suspected brain diseases. Thorotrast contains thorium which emits alpha particles, and is particularly likely to cause cancer of the liver. BEIR (1980, pp. 372-376) analysed follow-up studies of thorotrast patients in Germany, Denmark and Portugal. From about 3,800 patients there were 301 cases of liver cancer whereas only 6 would be expected. Hence, the injections of thorotrast raised the liver cancer rate by 50 times. It is estimated that the dose was about 25 rads (250 mGy) per year of alpha radiation to the liver. BEIR (1980) calculate that these results mean that if one million people were to receive one rad (ten mGy) of alpha radiation to the liver, 300 of them would develop cancer of the liver.

Van Kaick et al (1978) conducted a follow-up study of 2,108 West Germans who had been administered thorotrast and 1,815 controls (hospital patients of the same age and sex). They found that those who had received thorotrast had a rate of liver tumours that was over 50 times higher than the control group, whilst for myeloid leukaemia the increase was over 11 times. These results are in agreement with those reported by BEIR so far as the induction of liver cancer by thorotrast is concerned.

Mole (1978) has reviewed the studies of the thorotrast patients in Denmark, Portugal and West Germany and found that, in addition to causing a high rate of liver cancer, thorotrast also led to an increase in the leukaemia rate. Slightly over 1% of the patients exposed to thorotrast developed leukaemia, and it is estimated that the average annual dose to the bone marrow was 9 rads (90 mGy).

6.9 Phosphorus-32

Polycythemia vera is a blood disease which involves the over-production of red cells. In the late 1930's it was discovered that this disease could be controlled by repeated injections of phosphorus-32, which emits beta particles. These injections extended the median life expectancy of such patients to about 12 years after diagnosis. Mays (1973) estimated that the average skeletal dose received by those patients treated with phosphorus-32 was about 300 rads (3000 mGy). Of the patients treated with phosphorus-32 20% developed leukaemia whilst only 2% of the non-irradiated patients developed leukaemia. Mays (1973) calculated that the leukaemia risk was 60 leukaemias per year for every million person rads which is some 60 times higher than the corresponding figure for the Nagasaki bomb survivors.

6.10 Summary

Alice Stewart and others have established that the antenatal irradiation of children in the latter part of the pregnancy increases the rate of childhood cancer by about 50%. Since the amount of radiation involved is under 2 rads (20 mGy) this represents clear evidence of the dangers of even very low level radiation and has led to considerable caution in X-raying women who are thought to be pregnant. Various studies using the Tri-State Survey data have found evidence of concave dose-response curves that do not have a threshold dose. Rosalie Bertell has found that exposure to 0.6 rads (6 mGy) is equivalent to being one year older so far as the risk of leukaemia is concerned. Irwin Bross and others have used the Tri-State

Survey data to show that there are groups of the population who are particularly susceptible to radiation induced leukaemia. Bross has also shown that low level radiation can induce heart disease as well as leukaemia. The low levels of radiation involved in screening for breast cancer have been found to kill more women than are saved. In view of such evidence the U.S. government in 1978 accepted that no woman under 50 years should be subject to mammography. Karl Morgan has argued that the diagnostic use of X-rays in the U.S. is excessive and could easily be greatly reduced. There is also some evidence to suggest that this conclusion applies to the U.K., although to a lesser extent.

The therapeutic use of low level radiation to the thyroid in the treatment of ringworm has been shown to lead to increased cancer rates amongst the children concerned. As a result this medical procedure is no longer used. Higher doses of radiation to the thyroid to shrink its size have been shown to increase the rate of thyroid cancer by 80 times, and in consequence this treatment is no longer in use. High levels of radiation to the skull as part of the treatment for ALL are thought to reduce the IQs of the children concerned. The use of thorotrast as an X-ray contrast medium has been shown to increase liver cancer rates by about 50 times and this practice was banned in 1955. Phosphorus-32, which is used in the treatment of a blood disease, increases the leukaemia rate by ten times. However, it is felt that the therapeutic benefits of phosphorus-32 outweigh this greatly increased leukaemia rate, so this treatment is still in use.

7 Radiation exposure from nuclear explosions

7.1 Introduction

There have been two main types of exposure to nuclear explosions. First, there was the involuntary exposure of the Japanese populations of Nagasaki and Hiroshima in August 1945, when two nuclear bombs were dropped on them. Many of the survivors of these nuclear explosions received substantial doses of radiation. Second, the post-war atmospheric testing of nuclear weapons led to the compulsory exposure of military personnel and the involuntary exposure of the local population to low levels of radiation. These tests also resulted in radioactive fallout throughout the world, thereby involuntarily exposing everyone in the world.

7.2 Japanese Atom Bomb Survivors

In late 1945 the US Army surgeons issued a statement that all deaths due to the radiation effects of the Hiroshima and Nagasaki bombs had already occurred. They maintained this nonsensical view for the next seven years, Bertell (1985, pp.143-144). In consequence, it appeared that little would be learned by studying the Japanese survivors. However, in 1950 some five years after the explosions, a national census in Japan identified 284,000 survivors of the bombs dropped on Hiroshima and Nagasaki. Of these, 109,000 were selected by the U.S. ABCC for the Life Span Study. The ABCC has been succeeded by the RERF in conducting follow-up studies on the atom bomb survivors. The selected survivors were studied carefully for their subsequent causes of death, and post mortems were performed in about 40% of the cases to confirm the original diagnosis. The data on the exposed survivors was compared with similar data for non-exposed individuals in Japan. This provided a major source of information on the late effects of radiation. However, there are a number of problems with the ABCC data which will be considered in this chapter:-

(a) most of the observations are for people exposed to high doses of radiation and there are not enough observations of people exposed to low doses to determine the risks of exposure to low levels of radiation.

(b) the radiation dose was a mixture of neutrons and gamma rays, and neutrons are thought to be roughly ten times as carcinogenic as gamma rays.

(c) the radiation dose of a survivor depended on his or her location, i.e. distance from the explosion and also on his or her shielding at the time of the explosion. These two criteria considerably complicate the calculation of the dose received by each survivor.

(d) the atomic explosions devastated living patterns, and those who were still alive five years later, when the survivors were identified, were a selected group and possibly fitter than those who had died.

After considering some of the estimates of the dose-response curve based upon the ABCC data, criticisms of the ABCC data and studies based upon it will be discussed.

7.2.1 Leukaemia

Pochin (1976a) reported the results of a study by Moriyama and Kato in 1973. For each city they calculated the number of deaths from leukaemia during the period 1950 to 1972 for those survivors (males and females of all ages) receiving an estimated dose of between 0 and 9 rads (90 mGy) from the relevant bomb. These deaths were divided by the number that would have occurred if the death rate from leukaemia had been the same as in Japan as a whole. The resulting ratios were Hiroshima, 1.48 and Nagasaki, 1.78; each being significantly greater than one. This indicates that the low levels of bomb radiation of under 9 rads (90 mGy) led to an increase in leukaemia of 48% in Hiroshima and 78% in Nagasaki.

7.2.2 Concave Dose-Response Curves

A concave dose response curve is one that slopes up at a decreasing rate i.e. when viewed from below it is concave, see chapter 2. This means that the risk per rad is greater for low doses than for high doses. Baum (1973) summarized three studies of dose-response curve for Japanese survivors. He reports that, if dose response-curves of the form :- Response = (Dose in rads)^x are fitted to the data, the parameter x is usually estimated to be *less* than one. This implies that the relevant dose-response curve is concave, and linearity only occurs if x equals 1. Hence low doses were found to be proportionately more carcinogenic than high doses. The eight dose-response curves discussed by Baum (1973) are summarized in table 7.1 where it can be seen that seven of the estimated values of x are appreciably less than one. It should be noted that these dose response curves were fitted for survivors receiving between 1 and over 1,000 rads (10 and over 10,000 mGy), so concavity of the dose-response curve appears to apply over a wide range of doses.

Table 7.1 Summary of Three Studies of Cancer in Japanese Survivors

Cancer	City	x	Years	Author
All Malignancies	Hiroshima	0.50	1957-58	Harada & Ishida
Acute Leukaemia	Nagasaki	0.80	1946-65	Tomonaga
All Leukaemias	Hiroshima	0.66	1946-65	Tomonaga
All Leakaemias	Hiroshima	0.80	1950-66	Ishimaru et al
All Leukaemias	Nagasaki	1.00	1950-66	Ishimaru et al
Lung Cancer	Both Cities	0.19	1950-66	Ishimaru et al
Stomach Cancer	Both Cities	0.35	1950-66	Ishimaru et al
Female Breast Cancer	Both Cities	0.50	1958-66	Ishimaru et al

Rotblat (1978) also reports concave dose response curves for the Japanese survivors, as table 7.2 clearly shows.

Table 7.2 Excess Mortality per Million People per Rad

Rads	Breast Cancer Nagasaki	Lung Cancer Hiroshima
10 to 49	86	70
50 to 99	26	24
100 to 199	14	21
200 +	1	8

Table 7.2 shows that, for lower doses of radiation, the risk of breast cancer per rad is 86 times larger than for high doses of radiation, whilst for lung cancer the risk per rad from a lower dose is nine times larger than for high doses of radiation. Gofman (1983, pp. 228-229) has reanalysed the Japanese data for both cities combined and shown that the dose-response curves for both breast cancer and for all cancers are concave.

7.2.3 Neutrons, Hiroshima and Concave · Dose-Response Curves

The radiation from the Hiroshima bomb was previously thought to have been between one fifth and one third neutrons, whereas at Nagasaki there was no significant neutron exposure. The RBE of neutrons is estimated to be ten, i.e. one rad (10 mGy) of neutron radiation is ten times more carcinogenic than one rad (10 mGy) of gamma or X-rays. Therefore, in estimating dose response curves some allowance must be made for the type of radiation involved and this requires the use of an estimate of the RBE of neutrons. Mole (1975b) has pointed out large differences in the incidence of chronic granulocytic and acute other types of leukaemia between survivors at the two cites, with Hiroshima rates being roughly four times higher. He attributed this to a difference in the neutron content of the radiation between the two cities. For Hiroshima he fitted a dose response curve for these neutron associated leukaemias and the estimated neutron dose only for data for the years 1950 to 1966, and found that it was concave. This curve was significantly non-linear at the 3% level of statistical significance. Whilst not requiring an estimate of the RBE of neutrons, Mole has found further evidence to support the view that dose-response curves are concave. Subsequently, the neutron dose at Hiroshima has been revised downwards, see chapter 5.

7.2.4 Control Groups

As mentioned in chapter 4, any epidemiological study faces the difficult problem of finding a suitable control group. This is particularly true in the case of the Japanese bomb survivors. Four possibilities for use as a control group were considered by BEIR (1980, page 155) and all are open to criticism. The four possible control groups are:-

(a) *Not-in-city at the time of the explosion* . However, this group contains immigrants, who have a different underlying cancer rate, and some local people who, when they returned home, may have received a dose of low level radiation. For example, the neutron dose induced radioactivity in various stable chemical elements and irradiated those who entered the city shortly after the explosion, Pochin (1983, page 37).

(b) *The zero-rads group*. Whilst this includes 48% of the Hiroshima survivors it covers only 23% of the Nagasaki survivors. In addition it may be argued that the dose received by this group has been underestimated. In addition to radiation from various chemical elements, soon after the explosions there was 'black rainfall' in both cities. This rain contained radioactive fallout from the bombs, and many years later the local soil and farm produce in Nagasaki had an increased level of radioactivity,

Pochin (1983, page 37). Schmitz-Feuerhake et al (1983) have argued that the contribution of fallout must not be neglected. Gofman (1983, page 151) has estimated that the 'zero dose' group in fact received an average dose of 5.6 rads (56 mGy).

(c) *The zero to nine rads group.* At Hiroshima 22% of the survivors fall into this group, whilst for Nagasaki the corresponding figure is 33%. The average doses were 3.7 and 3.9 rads (37 and 39 mGy) respectively, whilst Gofman (1983, page 151) has estimated the dose to the 'zero to nine rads' group was an average of 2.2 rads (22 mGy). However, it is argued throughout this book that such low level doses cause a significant increase in the cancer rate. Therefore, if this group is used as the control, the risks will be understated. In addition, Schmitz-Feuerhake et al (1983) have estimated that the average internal dose received by this group was between 4 and 25 rads (40 and 250 mGy).

(d) *National cancer rates, adjusted for age and sex.* Because the cancer rates for any country will show considerable variation from area to area, the nation is not ideal for use as a control group.

Hence, there is no ideal control group for the Japanese bomb survivors. Studies relying upon this data must be interpreted in the light of this problem.

7.2.5 Biased Sample

The Japanese bomb survivors are unique, in that not only were they subject to a dose of radiation, but at the same time they were subject to injuries from the blast and burns from the heat flash. They then faced the psychological trauma of bereavements and the collapse of society. According to Rotblat (1977) under these conditions only the fittest managed to survive.

The Japanese bomb survivors used in the follow-up studies of the effects of radiation were not identified until five years after the explosions. Therefore the study population consisted of long term survivors or individuals who had either recovered from or avoided the earlier effects of blast and radiation, Stewart (1978b). Consequently, the long-term survivors were biased in favour of those whose initial potential for survival was above average. The implication is that those bomb victims who survived for over five years were less sensitive to radiation than the general population, so that estimates of cancer risks based upon Japanese survivors will be underestimates. Rotblat (1978) shows that the cancer risks for Japanese survivors are much lower per rem than for other populations, and so the Japanese data may be unsuitable for calculating radiation risk factors for other populations. Rotblat (1978) concludes that radiation is five times more dangerous than the current official risk estimates suggest, and proposes that existing dose limits be reduced to one fifth of their current levels. This view is reaffirmed in Rotblat (1982). After discussing likely biases in the Japanese sample, Stewart (1982) concludes that the official risk estimates are ten times too low.

7.2.6 Incorrect Dose Estimates

Recently the estimates of the radiation doses from the Japanese bombs have been questioned. It has been suggested that the neutron dose at Hiroshima has been overestimated by 600% to 2000% and at Nagasaki by 300% to 500%, whilst the gamma dose at Hiroshima and Nagasaki has been underestimated by 30% to 100% in each case, see chapter 5. Charles et al (1981) presented these doubts about the radiation doses received by the atomic bomb survivors and went on to derive risk estimates from data which excluded the Japanese bomb data. They concluded that, for low LET radiation, the risks are about 100% higher than official estimates based largely on the Japanese bomb data. So far as high LET radiation is concerned, if the Japanese bomb data is not used, Charles et al (1981) were unable to produce any estimates of the cancer risks. They stated that this highlights the paucity of data on the health effects of high LET radiation.

7.3 Atmospheric Tests

In the early 1960's it was realized that radiactive fallout from the atmospheric testing of nuclear weapons was leading to a build up of low level radiation throughout the world. In consequence in July 1963 the U.S., U.K. and Soviet Union agreed to ban further atmospheric nuclear tests. In an address to the American nation in support of the ban President Kennedy referred to the threat of fallout as follows: "The loss of even one human life, or the malformation of even one baby - who may be born long after we are gone - should be of concern to us all. Our children and grandchildren are not merely statistics toward which we can be indifferent," quoted in Sternglass (1981, page 28). Gofman (1976a) estimated that approximately one million deaths from lung cancer have occurred in the Northern Hemisphere as a result of plutonium fallout caused by weapons testing. In his Nobel Lecture in 1963 Linus Pauling stated:

> "It is my opinion that bomb-test strontium 90 can cause leukaemia and bone cancer, iodine 131 can cause cancer of the thyroid, and caesium 137 and carbon 14 can cause these and other diseases. I make a rough estimate that because of the somatic effect of these radioactive substances that now pollute the earth about two million human beings now living will die five or ten or fifteen years earlier than if the nuclear tests had not been made."

quoted in Tamplin et al (1970).

The atmospheric test ban treaty is one of the very few international agreements concerning nuclear weapons, and illustrates the importance which the governments concerned attached to the dangerous build-up of radioactive fallout which was recognised to be occurring in the early 1960's. In view of the estimates of Gofman and Pauling that atmospheric tests caused millions of deaths worldwide, it is not surprising that political leaders took extraordinary action to prevent the deaths of additional millions of people. After 1963 the signatories to the test ban treaty used underground tests which do not generate fallout and so are much safer; consequently the studies of the health effects of nuclear tests deal only with atmospheric testing. However, underground tests are not risk-free. Since 1963 the U.S. has conducted over 400 underground tests, some of which have leaked large amounts of radioactive gases, Bertell (1985, pp. 21, 56 and 163). These tests have also produced radioactive material in the ground which can contaminate local drinking water. This radioactive contamination of the environment by underground nuclear tests is unnecessary. In May 1985, Norris Bradbury, a former director of the U.S. Los Alamos weapons laboratory, wrote that 'continued nuclear testing is not necessary in order to insure the reliability of the nuclear weapons in our stockpile', quoted from Pirani (1986).

7.3.1 Smoky

Smoky, a 44 kiloton bomb, was detonated on top of a 210 meter steel tower adjacent to the Smoky foothills in Nevada on 31 August 1957. The explosion was witnessed by over three thousand toops from a distance of 29 kilometers. In addition to being present at the detonation of Smoky, many of the troops observed at least one other blast and practised manoeuvres three days later in an area possibly contaminated with fallout from Smoky.

Caldwell et al (1980) have studied 3,224 men who were present at the Smoky test. The film badge radiation dose is available for 3,153 of the men and the average radiation dose recorded for those with badges was 0.47 rems (4.7 mSv), whilst that for those who subsequently developed leukaemia was 1.17 rems (11.7 mSv). These film badge readings are cumulative for the entire series of 1957 tests and so include the effects of any other tests the man attended. The badges recorded only external radiation from gamma and beta radiation and did not measure possible internal radiation or radiation from neutrons. The observed number of leukaemia cases was 160% higher than expected on the basis of U.S. rates for 1970. These results will tend to understate the increase in the leukaemia rate for two reasons: (a) the control group was the U.S. population and so there will be a 'healthy worker effect' and (b) the follow-up period was twenty years and a few leukaemias may have appeared later. A later study by Caldwell et al (1983) revised the estimated increase in leukaemia incidence to 150% and the increase in leukaemia deaths as 158%. Current official estimates of the carcinogenic effects of one rem (ten mSv) suggest an increase in the leukaemia rate of about 30%, so the Smoky study indicates that, for radiation doses as low as only one rem (ten mSv), the cancer rate is five times higher than the current official estimates.

7.3.2 Nevada Tests

Between January 1951 and the end of October 1958 at least 97 atomic devices were detonated above ground in the Nevada desert at Yucca Flat, and fallout from at least 26 of these tests was carried by winds into Utah. On 11th January 1951 the AEC placed an advertisement in the local St. George newspaper. It stated that there would be no advance notice of the tests and that 'health and safety authorities have determined that no damage from or as a result of AEC test activities may be expected outside the limits of the (test) range', quoted from Reed (1984). Subsequent events have shown this public reassurance by the AEC to have been over-optimistic.

A classified US document that was obtained in 1980 under the Freedom of Information Act describes the radiation effects of a nuclear bomb test in Nevada in 1953. It states that the radiation intensity was very high and even 250 miles from ground zero the dose was 5 rads (50 mGy). A service station west of Bunkerville was subject to a dose of over 5 rads (50 mGy). The fallout was so great on the highways near Glendale that several vehicles were found to be highly contaminated. A considerable number of the personnel involved in this test were overexposed, Bertell (1985, page 167).

Lyon et al (1979) identified 17 rural southern and eastern counties of Utah as high fallout counties. The remaining urban centres of Utah, containing 90% of the state's population, were designated as low fallout counties since the fallout bearing winds were generally to the south. They studied the deaths from leukaemia of children aged between 0 and 14 years for these two areas, and also for the U.S. as a whole. Lyon et al did not present any numerical results, and their graphical results are plotted in figure 7.1 for the period 1944-1975.

Figure 7.1 shows that deaths from leukaemia for the high fallout area more than doubled for the period 1959-1967 returning to its previous level for the 1968-1975 period. For the low fallout area and the U.S. as a whole death rates from leukaemia showed little change for any period or sub-period. Inspection of the ages of the children dying of leukaemia in the high fallout area reveals the greatest excess to be in the 10 to 14 years group, i.e. those born around 1951 when nuclear testing commenced, and little excess in those born around 1961 i.e. after atmospheric testing had

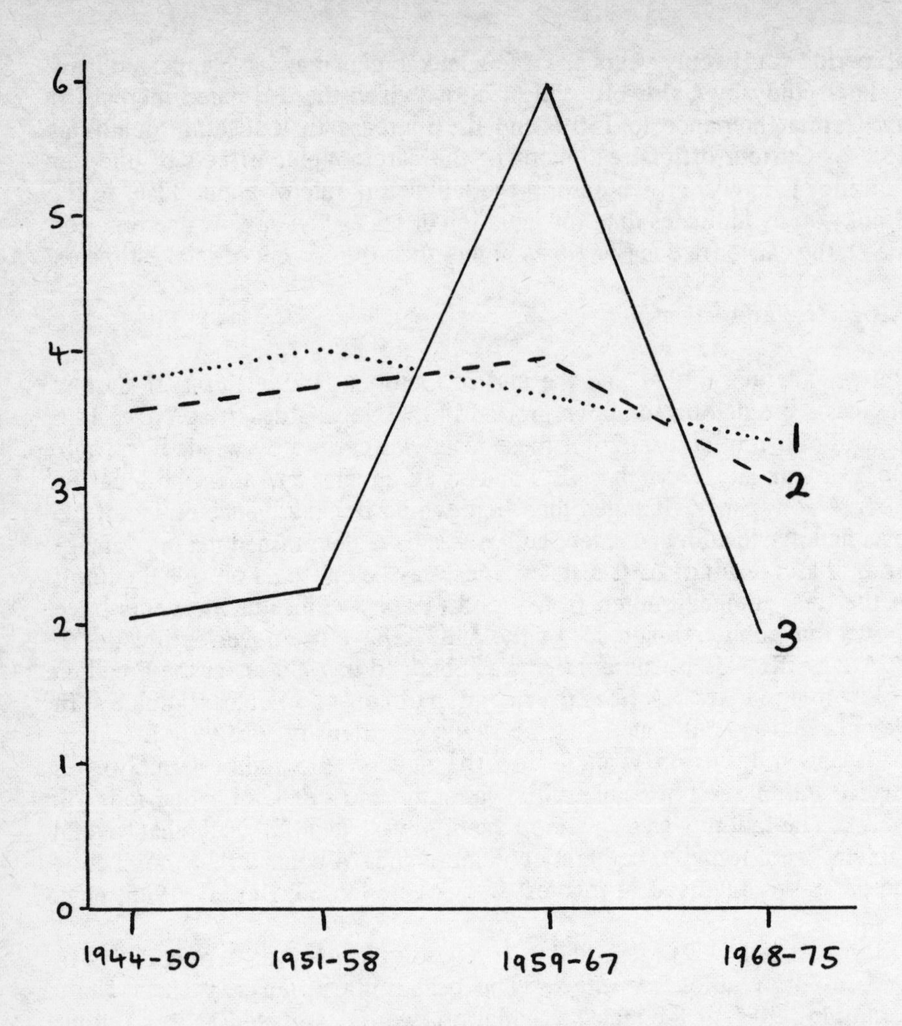

Figure 7.1 Childhood Leukaemia for High and Low Fallout Areas of Utah

1 = Low fallout counties, 2 = US, 3 = High fallout area

ended. No data is available on the doses of radiation from fallout received by the children concerned because no such measurements were made at the time.

According to Torrey (1980a) several studies are currently being undertaken by Dr. Edward Haskell of the University of Utah Medical Centre to try and reconstruct the actual radiation doses. Samples of brick that were fired before, during and after the period of atmospheric testing will be analysed using thermo-luminescent analysis which reveals the external radiation to which each brick has been subjected since firing.

Internal radiation will be estimated from post mortems of thyroid tissue of those who died during the tests, and lung tissue of those who died up to a decade after the testing (since plutonium will stay in the lungs for ten years). Torrey (1980a) states that Haskell thinks the highest doses of radiation will turn out to be between 3 and 5 rads (30 and 50 mGy).

In the spring of 1953, 4,200 sheep that had been spending the winter 80 km north of the nuclear test area at Yucca Flat were found dead. When the angry ranchers tried to get compensation, the government said that all the sheep had died from natural causes and the judge, Sherman Christensen, agreed, Torrey (1979). In 1979 Utah's Governor, Scott M. Matheson, searched the state's archives and found a 1953 Public Health Service post mortem report that some of the dead sheep had a concentration of iodine-131 in their thyroid glands that was one thousand times larger than the maximum permissable level for humans. In 1982 the judge who had ruled in 1956 that the sheep had died from natural causes ordered a retrial. In so doing Sherman Christensen said that the US Justice Department and the AEC were 'intentionally deceptive', Financial Times (1982). He ruled that the U.S. government had practised fraud upon his court in the 1956 trial. He said that the agents of the government had made false or misleading representations, that witnesses were improperly pressured and that crucial information was intentionally withheld or misrepresented, Smith (1982) and Barlett et al (1985, pp.310-314). The aim of the government deception was to prevent the sheep deaths from being linked to the nearby nuclear weapons tests.

The U.S. Public Health Service study was released in the summer of 1979 as a consequence of a lawsuit brought under the Freedom of Information Act. Torrey (1979, 1980a) states that this previously secret study, which was prepared by Edward Weiss in 1965, found a 47% increase in leukaemia in southern Utah between 1951 and 1964 and attributes this increase in leukaemia to the fallout from nuclear tests. Evidence such as this is being used by the residents of Nevada, Utah and Arizona in law suits against the U.S. government. 965 people are claiming damages of $2 billion in the largest single civil action ever brought by U.S. citizens against the federal government. Bertell (1985, page 37) reports that on 10th May 1984, District Court Judge Bruce S. Jenkins ruled on the first 24 claims. He awarded $2.66 million in damages to nine claimants, Barlett et al (1985, pp. 321-322).

The town of St. George, Utah lies about 145 miles east of Yucca Flat. The residents of St. George have been exposed to more fallout than has been recorded for any other populated area of the U.S. Panati et al (1982, page 174) reports that in 1954 shooting began in the St. George area on a film about Ghenghis Kahn called *The Conqueror*. The cast and crew spent 13 weeks filming in the zone contaminated by fallout. Subsequently one third of the cast and crew developed some form of cancer. Among the dead are the film stars John Wayne, Susan Hayward and Agnes Moorehead and the producer-director, Dick Powell.

Seventy two percent of the Utah population are Mormons and, because of their different lifestyle, they have a cancer incidence that is 16% lower than that of non-Mormons living in Utah. To control for religion Johnson (1984) restricted his study to Mormons. He investigated the rate of cancer incidence amongst Mormons living in the high fallout area in the south west of Utah. This high fallout area covered the towns of St. George, Parowan, Paragonam and Kanab in Utah and also Fredonia in Arizona and Bunkerville in Nevada, see figure 7.2 The exposed group was defined as the Mormon families living in the towns between 1951 and 1962. In 1981 Johnson managed to trace 4,125 such people. The control group comprised all Mormons living in Utah over the period 1967 to 1975, adjusted for differences in age and sex. (In 1971 there were 781,735 Mormons living in Utah.) The increase in incidence for radiosensitive cancers for two time periods is given in table 7.3. This shows that in each period the cancer rate for radiosensitive organs roughly doubled. For each period the increase was statistically significant at the 1% level.

Table 7.3 Increases in the Rate of Cancer Incidence for Radiosensitive Organs.

Years	High Fallout Area	Fallout Effects Sub-Group
1958-1966	82%	567%
1972-1980	104%	567%

There are a number of factors which will tend to lead to the understatement of any excess cancers by this study:-

(a) the control group includes most of the exposed group,

(b) all the residents of Utah were exposed to some fallout from nuclear weapons tests, so the cancer rate for the control group will be higher than otherwise,

(c) about 40% of the exposed group could not be located in 1981, and this was thought to lead to an understatement of the risks,

(d) whilst more Utah Mormons live in urban areas, the fallout area is rural, and this would tend to produce lower cancer rates in this area.

Therefore, Johnson's finding that cancer incidence in the high fallout area roughly doubled, may be an understatement of the effects of the nuclear weapons testing on local cancer rates.

Johnson also identified a sub-group of 239 persons from the high fallout area who had experienced acute fallout effects. All Utah Mormons were again used as the control group. The results in table 7.3 show that the incidence of radiosensitive cancers amongst this fallout effects subgroup increased by almost 600%. Again the increases in cancer incidence were statistically significant at the 1% level.

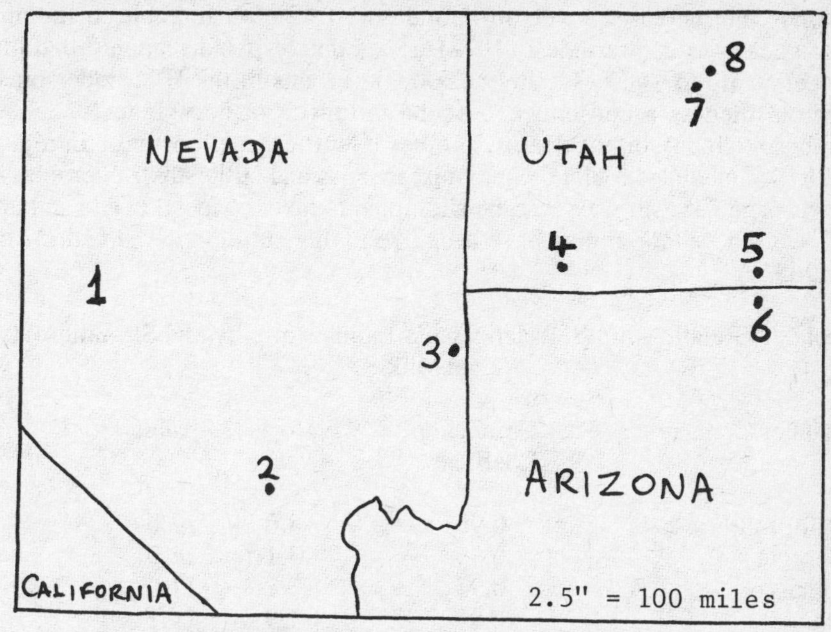

Figure 7.2 Towns Downwind of Yucca Flat

1 = Yucca Flat, 2 = Las Vegas, 3 = Bunkerville, 4 = St. George, 5 = Kanab, 6 = Fredonia, 7 = Parowan, 8 = Paragonan

7.3.3 Strontium-90 in Milk and Infant Mortality

The atmospheric tests produced fallout containing strontium-90 which was then ingested by cows leading to the contamination of milk. In the late 1960's global fallout caused young children to have one and a half to two times the adult levels of strontium-90 in their bones, Pochin (1983, page 76). Not only can radioactivity cause cancer, it can also adversely affect foetal growth, cause miscarriages, stillbirths, and reduce birthweight; it is also associated with increased deaths during the first year of life. These effects follow within months of the release of the strongtium-90. For nine U.S. states and also for the U.S., U.K. and New Zealand, Sternglass (1972) calculated the excess infant mortality rates over the period from about 1957 to 1971. This was done by extrapolating the past trend in infant mortality rates and then calculating the extent to which the actual mortality rates exceeded the predicted rates. Sternglass then regressed these excess infant mortality rates on the levels of strontium-90 in local milk using annual data. His results are summarized in table 7.4. This shows that for all the areas considered there is a highly significant positive effect of strontium-90 on infant mortality.

Since the increased levels of strontium-90 were attributable to fallout from nuclear tests, Sternglass (1972) argued that his results suggest that in the period up to 1965 as many as 400,000 infants in the U.S. under one year old died as a consequence of the fallout from nuclear tests. This number excludes infant deaths in other countries, e.g. northern Europe, and it also excludes foetal deaths (miscarriages and stillbirths). Sternglass (1981, page 73) estimates that the additional number of foetal deaths in the U.S. caused by fallout may have been two to three million in the period up to 1965.

Table 7.4 Relationship Between Excess Infant Mortality and Strontium-90 in Milk

State or Country	Correlation Coefficient	Slope	t value
California	0.96	8.0	12.0
Georgia	0.95	3.1	11.0
Illinois	0.95	3.7	11.0
Ohio	0.98	1.9	14.1
Missouri	0.97	4.0	13.4
New York	0.97	3.5	13.0
Texas	0.97	4.6	13.2
Utah	0.84	3.2	5.9
Washington	0.91	1.1	7.3
U.S.A	0.98	3.2	18.3
England and Wales	0.92	3.7	7.5
New Zealand	0.95	3.8	9.2

7.3.4 Nassau County

Sternglass (1972 and 1974) studied the relationship between excess infant mortality in Nassau County, New York and fallout as measured by the external gamma radiation dose at the Brookhaven National Laboratory, Suffolk County. Excess infant mortality was calculated by drawing a line through the infant mortality rate in 1956 and 1966. Deviations above this line were then used as the excess infant mortality figures. Sternglass regressed annual excess infant mortality for Nassau county on the annual fallout figures. The correlation coefficient was 0.80 whilst the slope was 0.22 with a t value of 4.2. This indicates that fallout has a highly significant positive effect on infant mortality, while an increase of radiation dose of as little as 0.1 rad (1 mGy) per year will cause the infant mortality rate to be 22% higher.

Sternglass (1972 and 1974) also examined the connection between levels of fallout in Nassau county and leukaemia. Since leukaemia has a latent period Sternglass regressed the percentage increase in leukaemia in Nassau county on the level of fallout five years earlier. The correlation coefficient was 0.82, the slope was 0.49 and the t value was 3.5. Thus fallout has a statistically significant effect on leukaemia, with an increase of 0.1 rad (1 mGy) per year leading to a 49% increase in leukaemia.

7.3.5 Fallout and U.S. Aptitude Scores

In 1975 the scores in the U.S. nationwide Scholastic Aptitude Test (SAT) dropped by the largest amount in two decades. Since most of those taking the test were aged 18 they were born in 1957, the year when radioactive fallout in the U.S. was at its height. Sternglass (1981, pp. 181-191) hypothesised that those children who had not been killed by the fallout in 1957 may have had their development retarded and their intellectual performance reduced. The dramatic drop in SAT scores led to the establishment of a special panel to investigate the causes, but despite commissioning more than 24 special research studies this panel was unable to find any cause. Sternglass then analysed the SAT scores on a regional basis since, according to his hypothesis, those regions subjected to the highest fallout levels should have had the largest drop in SAT scores, and this is what he found. The other prediction of the hypothesis of Sternglass was also confirmed when, as the U.S. fallout levels at the time of birth decreased, so the SAT scores recovered.

7.3.6 Marshall Islanders

On 1st March 1954 the nuclear bomb 'Bravo' was detonated at Bikini atoll vaporizing the small island on which the explosion occurred. This was the largest ever U.S. atmospheric test producing a blast 750 times larger than the Hiroshima bomb. There was a massive plume of powdered coral and radioactive material extending hundreds of miles downwind of the explosion. Part of this radioactive material fell on a number of the Marshall Islands including Rongelap, which is about 100 miles east of the Bikini atoll. The U.S. government has said that it was an accident that the wind blew the fallout of the Bravo explosion towards the Marshall Islands. Gene Curbow was a member of the unit that measured wind velocity for the Bravo test. He has said that 'for weeks prior to the blast it was known that prevailing upper level troughs indicated that the winds were blowing in the direction of inhabited islands', Saffer et al (1982, page 213). The radioactive ash on Rongelap, which was between two and three inches deep, was played in by the local children, Alcalay (1980). For a map of the Marshall Islands see figure 7.3.

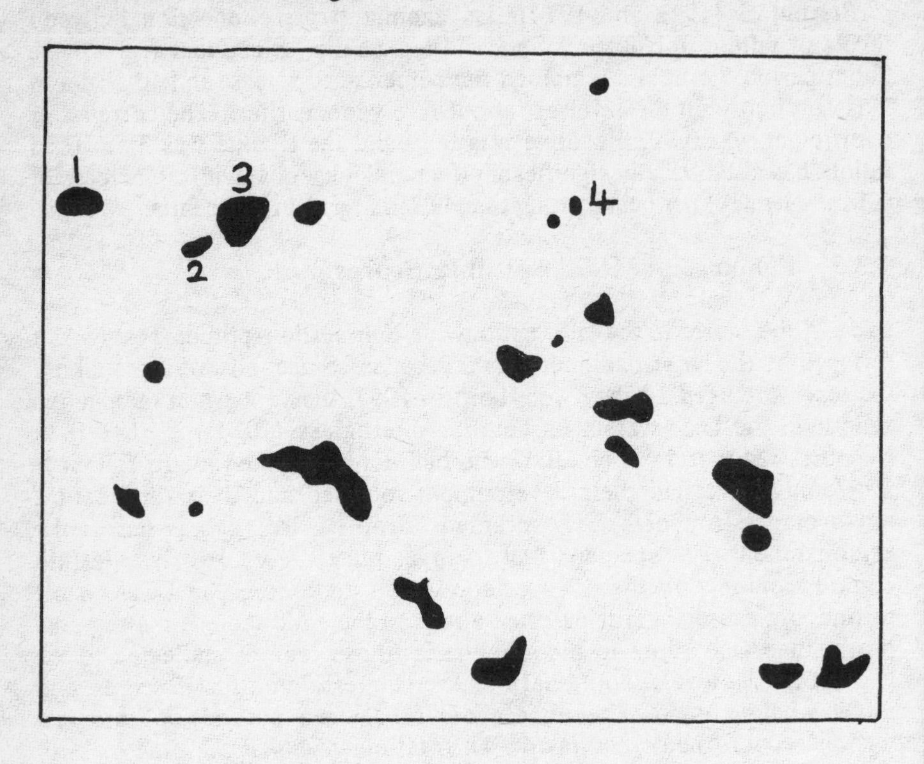

Figure 7.3 The Marshall Islands

1 = Bikini, 2 = Ailingnae, 3 = Rongelap, 4 = Utirik

The islands of Rongelap, Ailingnae and Utirik were evacuated about 48 hours after the fallout first appeared. Larsen et al (1978) note that, following the explosion, the population of Rongelap exhibited the acute effects of gamma radiation exposure. These were a loss of appetite, nausea, vomiting, reduction of white cells by half, radiation burns and loss of patches of hair. There was also an increase in miscarriages and stillbirths among the exposed Rongelap women. Mays (1973) reports that a large part of the dose received by the Rongelap islanders was from inhaling radioactive iodine and drinking water contaminated with radioactive iodine. A follow-up study by Conrad (1977) estimated that the average dose to the thyroids of children aged under 10 years from both internal and external radiation was about 1,010 rads (10,100 mGy).

By 1976, 22 years after the explosion, 18 of the 23 Rongelap children aged under 10 at the time of the explosion had developed thyroid tumours, i.e. 78%. (Using figures up to September 1977 Larsen et al put this figure at 90%.) This compares with a corresponding figure of 1.6% for a control group of Rongelap children who were away from the island at the time of the accident and who subsequently moved back to the island. Conard (1977) points out that the adverse health effects recorded so far may understate the risks for two reasons. New thyroid tumours are still developing so that for a longer follow-up period a still greater effect may be found, and the hormone and surgical treatment which the islanders have received may have helped to reduce the measured adverse effects. Larsen et al (1978) conclude that the death from leukaemia of a Rongelap man who was one year old at the time of the explosion was probably due to the nuclear explosion, whilst five children who were exposed at less than five years of age have suffered growth retardation. Two children have developed hypothyroidism and Larsen et al (1978) suspected that other islanders who were exposed to the radiation may also be developing thyroid malfunctions. Because of this they conducted a test involving a thyroid stimulating hormone which is the most sensitive index of impaired thyroid function. They tested 43 people who were exposed to radiation on the islands of Rongelap and Ailingnae and who had no known thyroid abnormalities. As a control they also tested 115 Marshall Islanders who were not exposed to the radiation. At the 5% level of statistical significance they found evidence of early thyroid disfunction.

Not only did the Bikini tests accidentally irradiate the Marshall Islanders with adverse health consequences, these tests also had adverse consequences for the population of Bikini Island due to a mistake by the U.S. government. In the New Scientist (1981) it is reported that in 1966, based upon a 1957 study, the U.S. government decided it was safe for the Bikini Islanders to return home. But in the mid 1970's the U.S. government realized the levels of radiation on the island were still too high for anyone to live there. A survey in 1978 estimated the island would be too dangerous for habitation for another 30 to 60 years, and the islanders were again evacuated. In 1978 the Bikini Islanders started a law suit against the U.S. government claiming that they have ingested more radiation than any other population in the world. In May 1985 the entire population of Rongelap (304 people) was evacuated by the Greenpeace ship Rainbow Warrier, to an uninhabited Pacific island, Brown (1985). This was because the Rongalese have decided that nuclear weapons fallout has made their island uninhabitable, and appeals to the U.S. government have been ignored.

As a sergeant in the U.S. army stationed on Eniwetok Island in the Marshalls, Orville Kelly was ordered to witness twenty two nuclear explosions between 10th April 1958 and 31st August 1958. The U.S.

army estimated that during this period Kelly received a radiation dose of 3.5 rem (35 mSv). In 1974, at the age of 43, he was found to have cancer and he then sought compensation from the U.S. government. In November 1979 Kelly became the only U.S. serviceman so far to obtain a disability benefit because of a cancer caused by witnessing nuclear weapons tests. Seven months later, in June 1980, he died of cancer, Saffer et al (1982).

7.3.7 Japanese Fishing Vessel

At the time of the Bikini explosion the Fukuryu Maru (the Lucky Dragon), a Japanese fishing vessel, was eighty miles east of the atoll. The boat was covered in fallout from the Bikini bomb, and it is estimated that the twenty three crew members received an external whole body dose of between 270 and 440 rads (2,700 and 4,400 mGy), Nishwaki (1954). In addition, they will have received an internal dose from breathing in dust, eating contaminated food and drinking contaminated water. A dose of between three and four hundred rads (3,000 and 4,000 mGy) will kill half those exposed within thirty days of the exposure i.e. LD50-30 (the lethal dose for fifty per cent of people within thirty days), Pochin (1983, pp.102-103). The crew suffered from severe radiation burns and loss of hair. On their return to Japan they were hospitalized and received blood transfusions. They were seriously injured and Nishwaki wrote that 'the condition of some of them, those most seriously injured, does not seem to warrant any optimism'.

The tuna fish caught by the boat were found to be emitting up to five hundred times the normal level of beta radiation. Unfortunately, some of this fish was sold at Osaka City Central market and consumed by about one hundred people.

So, in addition to accidentally irradiating the Marshall Islanders, the Bikini bomb also irradiated the 23 people on a Japanese fishing vessel. The crew suffered from the early effects of radiation exposure because of the size of dose to which they were exposed. They are also subject to a greatly increased risk of cancer.

7.3.8 South Pacific

Between May 1957 and November 1958 the UK conducted ten atmospheric nuclear tests in the South Pacific, whose local base was Christmas Island. These tests were witnessed by British military personnel. In December 1982 - January 1983 the BBC TV programme *Nationwide* broadcast the experiences of some of the servicemen and their subsequent ill health. This produced a flood of letters concerning other servicemen who were present at the tests but had not been contacted by the programme's researchers. Many of the servicemen did not wear radiation badges so the size of the radiation dose they received is unknown, but some men were knocked flat by the blast which also bent over palm trees, whilst the flash dried out the men's clothing. The servicemen had been told just to lie down with their backs to the blast and their hands over their eyes.

Colin Avey was a 17 year old seaman on HMS Scarborough. The only precaution that was taken by the crew was that they had to change into winter trousers and long sleeved shirts. Just before the explosion they were ordered on deck to sit with their backs to the bomb and their hands over their eyes, Davies (1983). The Scarborough's job was to steam to ground zero measuring the atmospheric effects. Colin Avey subsequently developed leukaemia. At the time of an explosion Ken McGinley was standing on the beach of Christmas Island wearing only his tropical kit. He has said that 'all we were told to do was to turn round and keep our eyes covered against the flash', Stead (1984). There were several hundred men on the beach and the blast knocked them backwards by about four feet. Four days later Mr. McGinley broke out in water blisters on his face and chest. He returned from Christmas Island with internal bleeding and was admitted to hospital in the UK Colonel Peter Lowe observed a test from inside a tank. He must have been fairly close to ground zero because the blast moved the tank about ten feet sideways. He subsequently developed stomach cancer, Smith (1985, pp. 19-20). In October 1953, a Canberra bomber was flown through the nuclear plume, seconds after the bomb explosion at Emu Field in South Australia. One of the purposes of this flight was to test the effects of intense radiation fields on the aircraft and its crew, and instruments revealed that the crew were exposed to a dose of 22.8 rads (228 mGy), Daily Telegraph (1985). A secret UK government document written in 1953 stated that 'the Army must discover the detailed effects of various types of explosion on equipment, stores and men *with and without protection*', Smith (1985, page 110). In consequence men were placed as close as one and a half miles from ground zero.

The BBC received over 80 letters describing the subsequent ill health of servicemen who had been present at these tests. The BBC sent these letters to Dr. Alice Stewart, who had agreed to conduct a study of cancer incidence, and the results were published a few months later in Knox et al (1983). Information was obtained on only 330 servicemen who had been at the tests, of which 69 had died of cancer. These cancer deaths included 27 deaths from leukaemia and other RES cancers. In addition, thirteen living servicemen were found who are suffering from RES cancers. Using a Parliamentary answer by the Prime Minister on 8th February, 1983, the total number of servicemen involved in the tests was estimated at 8,000. For 8,000 servicemen the expected number of deaths from RES cancers in the absence of any exposure to radiation, is 17.2. Since 27 such deaths are known to have occurred, and assuming that *none* of the other 7,670 uncontacted servicemen has died from an RES cancer, the death rate from this cause is about 60% higher than expected. If any of the uncontacted servicemen have died from an RES cancer the increase in the death rate is larger than 60%. Knox et al also found that, of the 214 responses concerning men under 25 years of age in 1958, ten had cataracts. In the view of Boag et al (1983) 'the reported incidence of cataract, virtually unknown as a spontaneous occurrence among young men, is a strong indication that some of those involved had received radiation greatly in excess of a safe dose'.

Very shortly after the Knox et al study was published, the Ministry of Defence (MoD) challenged the use of 8,000 as the total number of exposed servicemen. Whilst the 8,000 figure was based upon a statement by the Prime Minister, the MoD has claimed that a number of about 12,000 should have been used, New Scientist (1983). If the MoD's figure is correct, the expected number of deaths from RES cancers rises from 17.2 to 25.8 and most the of excess found by Knox et al disappears (assuming that *none* of the 11,670 uncontacted servicemen has died from an RES cancer).

The MoD have received four claims from people seeking compensation for cancers allegedly caused by the tests. All have been turned down. The Crown Proceedings Act (1947) prevents British servicemen from suing the MoD for damages. However, the BBC found that in one case turned down by the MoD (a widow whose husband died of leukaemia in 1980) the claimant had been awarded a war widow's pension by the Department of Health and Social Security (DHSS) who have accepted that exposure to radiation caused by the Christmas Island tests was responsible for his death. In 1980 Mike Saffrey, an RAF radio operator on Christmas Island in 1958, applied to the DHSS for a war pension on the grounds that the radiation had caused his cataracts and subsequent blindness. His application was refused, but on 26th July 1984 his appeal against the DHSS decision was successful 'on the basis that the disability is attributable to service', Smith (1985, page 165). Whilst never having undertaken a survey of any kind, the MoD has claimed that the tests were *safe* and 'carrying out a full epidemiological survey would be a very large and costly task and would be nugatory', quoted from Tucker (1983a). Enquiry into the health effects of the tests by others has been hampered by the Official Secrets Act. But, following the adverse publicity generated by the TV programmes, the MoD has now agreed to an investigation into the health effects of these tests.

In addition to the Christmas Island Tests, there were twelve British nuclear tests in Australia between October 1952 and October 1957, with seven tests at Maralinga (an Aboriginal word meaning thunder), three at Montebello and two at Woomera, Tucker (1983b). There is also considerable concern about the health consequences of these tests, and the Australian Government set up a Royal Commission to investigate the effects.

At Maralinga servicemen went into the high blast areas shortly after the explosion to examine and remove equipment and vehicles which had been deliberately exposed. For example, Warrant Officer William Jones of the Australian army, then aged 26, was in charge of a Centurian tank which had been placed close to ground zero. Shortly after the explosion William Jones and his crew went into the blast area and tried to start the tank, but could not do so. He remained with his tank for some 48 hours whilst his crew went off for spares to repair the tank. Not only would the area round ground zero have been radioactive but, as a result of neutrons from the bomb, the tank itself will have become radioactive. William Jones subsequently died of cancer at the age of 39. In 1974 his widow, Peggy Jones, was subsequently awarded compensation from the Australian government of $8,000 and small weekly payments for each of her four children, Smith (1985, page 106).

Not only were servicemen irradiated by the Australian tests, but groups of aboriginies were also exposed to radiation. Because the aboriginies were used to freely wandering the desert area, they were difficult to control. It has been claimed that some aboriginies:-

(a) were inside the exclusion zone at the time of the explosions,

(b) were outside the exclusion zone but watched the explosions, so damaging their eyes,

(c) wandered into areas of high residual radioactivity after the explosions.

In 1984 John Burke, a former RAF technician dying of stomach cancer in Adelaide, revealed that after one of the British nuclear tests in 1963 he had found the bodies of four Aboriginies in the bomb crater, Bertell (1985, page 166) and Smith (1985, page 128). This supports the view that the British tests did kill Aboriginies.

It has been claimed that large amounts of fallout were produced by the tests in Australia and that this is shown by studies of the uptake of radioactive material by sheep, Tucker (1983b). However, these studies appear to have been largely suppressed. In 1984 granules of plutonium weighing up to 0.3 oz were found near Maralinga, Financial Times (1984). Evidence to the Australian Royal Commission has revealed that, in experiments at Maralinga designed to measure the radiological hazards from an accidental nuclear explosion, plutonium was burned in petrol fires and the plutonium oxide scattered downwind, Milliken (1984).

The report of this Commission (1985) was critical of the conduct of the nuclear weapons tests in Australia by Britain. It was concluded that inappropriate locations were chosen for some of the tests, bombs were denonated in unsuitable weather conditions that violated the rules, and some military personnel were not given protective clothing or proper instructions. There was a disregard for the safety of the aboriginies, and incidents occured which bureaucrats, scientists and politicians had given categorical assurances could not occur. The Commission also found that the conclusion of the 1983 report of the Australian Ionising Radiation Advisory Council, that the tests were conducted safely, was directly contrary to the available evidence.

7.3.9 Soviet and Chinese Nuclear Bomb Tests

Since 1949 the USSR has conducted nuclear bomb tests north of Lake Balkhash in Siberia. There are two towns within 62 miles of the test site, each with a population of about 100,000 people. Bertell (1985, page 165) states that there are no published reports of the effects of these tests upon the health of the local population. However, the risks to the health of the local population will presumably be of a similar nature to those found in the USA around the Nevada test site.

While the major nuclear powers ceased atmospheric testing in 1963, China exploded nuclear weapons in the atmosphere until 1983 in the Xinjiang region of north west China, MacDougall (1986). Various accidents have taken place, and it is probable that there have been adverse health effects.

7.4 Summary

The hostile and the peaceful use of nuclear weapons has been shown to have led to the involuntary exposure of humans with adverse consequences to their health. The Japanese data, which has been a major piece of evidence in the estimation of the health risks of radiation, was shown to yield concave dose-response curves i.e. greater risk at lower doses. However, the appropriateness of the Japanese data for calculating the risks of exposure to low levels of radiation has been seriously questioned, e.g. insufficient observations at low doses, difficulties and inaccuracies in calculating the dose, the mixture of gamma radiation and neutrons and the use of a sample that was selected five years after the explosions.

The signing of the nuclear test ban treaty shows that world leaders accepted the serious consequences of atmospheric testing for the world population. The subsequent health studies support the view that very low doses of radiation can have adverse effects on the health of populations involuntarily exposed. The accidental exposure of the Marshall Islanders and the ill-advised early return of the Bikini Islanders, shows that despite assurances to the contrary, governments have unwittingly exposed local populations to doses of radiation well above the safety standards.

8 Occupational exposure to radiation

8.1 Introduction

A variety of people are exposed to radiation as part of their jobs, and considerable anecdotal evidence is available of radiation workers who have subsequently died of cancer. However, for the reasons set out in chapter 4, this does not constitute proof that occupational exposure was responsible for their deaths. In any group of people there is a chance that some of them will die of cancer, even if none of them has been exposed to radiation. In order to provide firm evidence of adverse health effects, an epidemiological study is required, and such evidence is available for only a few types of occupational exposure. To carry out an epidemiological study, data is required on a large number of exposed workers over a period of several decades and, to the extent that such data exists, it is in the hands of the employers concerned e.g. government agencies, the military and private companies engaged in the nuclear industry. These employers do not wish to see studies conducted which prove that their employees are being killed by their occupation, particularly because this will expose the employer to expensive law suits from their workers for compensation. As has been pointed out in chapter 3, the researchers carrying out both the Hanford and Portsmouth studies of occupational exposure experienced considerable difficulties in obtaining their data.

The maximum permissible exposure of workers recommended by the ICRP (whose recommendations have been adopted by most countries) is 5 rems (50 mSv) per year. This is 10 times larger than the maximum permissible exposure for the general public of 0.5 rems (5 mSv) per year recommended by the ICRP. The reasons for this difference in exposure standards are not clear since radiation workers have no special immunity to health damage resulting from radiation exposure.

There is an argument that radiation workers have voluntarily chosen to expose themselves to radiation hazards and are compensated for taking these risks by higher rates of pay. This implies that no safety standards for occupational exposure are required, the levels of occupational exposure being settled by negotiation between the employer and the worker. However, such negotiations, with resulting differences in the exposure limits and wages applicable to different workers, do not in fact take place. The limitations on the radiation exposure of each worker are determined by regulation not negotiation. There is an alternative explanation for the tenfold difference in exposure standards between workers and the general public. To introduce a common exposure standard of 0.5 rems (5 mSv) per year would create very considerable problems for the continued functioning of many existing nuclear establishments and greatly increase their costs.

The current practice of "burning out" workers suggests that the existing exposure standards for workers are already too strict for some nuclear establishments. Bodanis (1982) describes how unemployed people are recruited by the U.S. nuclear industry to crawl into heavily contaminated nuclear steam generating chambers to seal pipes leaking radioactive material. In one five minute session these workers can receive a radiation dose equivalent to 100 chest X-rays, and each will be exposed for three or four five minute sessions. After less than an hour's work they have received a large proportion of the annual maximum permitted radiation dose and so cannot be employed again for some time. This practice of "burning out" workers means that jobs which have a very much higher annual rate of radiation than the permitted 5 rems (50 mSv) can be accomplished. It is estimated that in 1979 half of the occupational dose in the U.S. nuclear power industry was received by glowboys, Gofman (1983, page 353) whilst by 1990 the U.S. nuclear industry will need to "burn out" 84,000 people each year.

There is an ever increasing number of people who are exposed to radiation as part of their work. For example, it has been estimated by the Canadian Labour Congress (1984) that the number of radiation workers in Canada is expanding at the rate of about 10% per year, i.e. doubling every seven and a quarter years. Some examples of the type of jobs done by radiation workers are given by Pochin (1983, pp. 28-32 and page 78), and they show the very broad range of activities that now involve radiation. These are measuring the wear of mechanical components, checking welds, castings and pipelines, well logging, detecting art forgeries, monitoring the thickness of paper, sterilizing dressings, preserving food, pest control, causing mutations to produce resistant crops, phosphate fertilizers, diagnostic radiology and radiotherapy staff, workers at radon spas, aircrew, astronauts, watch liminizers, starters for flouorescent lamps, cardiac pacemaker batteries, lightship batteries, uranium miners, enrichment of uranium, manufacture of radioactive fuel elements, reactor operators, reactor maintenance staff, reprocessing, dockyard workers servicing nuclear submarines, military personnel and nuclear researchers (weapons, reactors, etc.). In 1986, the British government estimated that 354,000 workers were occupationally exposed to radiation, of which 135,000 have their radiation exposure monitored, Atom (1986a).

This chapter presents studies of three phases of the nuclear fuel cycle: uranium mining, and the operation of nuclear reactors and a reprocessing plant. There is also some evidence available on the adverse health effects resulting from the servicing of nuclear powered submarines and from the manufacture of nuclear weapons. In addition, there have been studies of the effects of the exposure of radiologists to X-rays and radium dial painters to radium. These are all considered in this chapter. The studies of servicemen exposed to radiation during the testing of nuclear bombs have been discussed in chapter 7 whilst the exposure of workers in radon spas will be considered in chapter 10.

8.2 The Hanford Study

The most important study of occupational exposure to low levels of radiation took place at the Hanford Works in Richmond, Washington, which is one of the largest atomic plants in the U.S. Nine nuclear reactors were built at Hanford for the sole purpose of producing plutonium for U.S. nuclear weapons. Since 1943 regular radiation monitoring of both the internal and external radiation received by each worker has been carried out. External radiation (largely gamma radiation) was measured by radiation badges, whilst internal radiation was measured by urine analysis. Using social security numbers it was possible to follow up the deaths of employees after they ceased employment at the Hanford Works, with the cause of death being taken from the death certificates. There have been a number of analyses of the Hanford data, with the later studies using data for a longer time period. After describing the first study which sparked off the controversy, subsequent re-analyses of the data will be considered. In view of the important and controversial nature of the Hanford study, the various analyses of the data will be described in some detail.

8.2.1 Mancuso, Stewart and Kneale (MSK)

The initial study was conducted by Mancuso, Stewart and Kneale (1977), and the background to this study was described in chapter 3. The results are summarized in Stewart(1978d). MSK used annual data for 29 years covering the period from 1944 (when the works went into full production) to 1972. A summary of their data appears in table 8.1.

Table 8.1 Certified Deaths 1944 - 1972

	Males		Females	
	Number	Average Dose*	Number	Average Dose*
Cancer deaths	670	1.38	127	1.33
Non-cancer deaths	2850	0.99	285	0.68
Total	3520	1.07	412	0.88

*Average cumulative dose of external radiation measured in rads.

They concluded that the radiation dose received by those workers who died of cancer was "appreciably higher" than for those who died from some other cause. They then went on to calculate for males the cumulative mean dose and the ratio between the observed and expected deaths for each category of cancer. The results for the six highest observed/expected ratios are set out in table 8.2.

Table 8.2 Male Deaths by Type of Cancer

Type of Cancer	Cumulative Mean Dose (rads)*	Observed/ Expected
Myeloid leukaemia	1.22	1.90
Myelomas	7.75	1.45
Liver and gall bladder	0.31	1.44
Kidney	1.87	1.40
Lung	1.69	1.33
Pancreas	2.53	1.31

*Average cumulative dose of external radiation measured in rads.

The expected values for table 8.2 were calculated from the 1960 cancer deaths of U.S. white males. These findings suggested a causal association between radiation exposure and cancer, but this could have been due to some other factor. Therefore, the authors controlled sepately for five other possible causative factors:-

(a) calendar exposure years

(b) employment exposure years
(c) pre-death exposure years
(d) exposure age
(e) death age

They conclude that their initial impression of a causal connection between exposure to external radiation and cancer deaths is strengthened rather than weakened by controlling for these five factors. Having accepted that there is a causative link between radiation and cancer deaths amongst the Hanford workers, the authors went on to estimate the doubling doses. The doubling dose is that dose of radiation which is required to double the normal risk of dying from the specified form of cancer. Hence the smaller is the doubling dose the more sensitive is that form of cancer to exposure to low levels of radition. The doubling doses which the authors estimated for males are set out in table 8.3.

Table 8.3 Doubling Doses for Males

Type of Cancer	Doubling Dose in rads*	Doubling Dose in mGy*
Bone marrow	0.8	8
All reticuloendothelial system (RES) neoplasms	2.5	25
Lung	6.1	61
Pancreas	7.4	74
All cancers	12.2	122

*Average cumulative dose of external radiation measured in rads (mGy).

The estimated doubling doses relate to cumulative radiation since employment commenced, and the maximum permissible occupational dose of gamma radiation is 5 rems (50 mSv) per year e.g. 5 rads (50 mGy) of X-rays. Therefore, table 8.3 shows that workers could quite legally receive a doubling dose for bone marrow cancers and all RES neoplasms each and every year, whilst they could also receive a doubling dose of radiation for lung cancer and cancer of the pancreas every two years. When they investigated the doubling doses by age, the authors found evidence that men aged under 25 or over 45 were especially sensitive to radiation caused cancer deaths.

An analysis of the 411 female deaths broadly supported the findings from the male deaths, but since the number of observations was considerably smaller, these results were not reported at length. There was some data on workers whose bodies contained radioactive material and so were subject to internal doses of radiation, but this data was not in a suitable form for further analysis. Finally, using their estimates of cancer risk due to radiation, the authors estimate that between 6% and 7% of all cancer deaths amongst Hanford workers were radiation induced. They calculated that roughly 26 Hanford workers died of cancer as a consequence of exposure to low levels of radiation in the course of their employment. These levels of radiation were well within existing safety standards.

Since their estimates of the doubling doses were so much lower than those based upon atomic bomb survivors, MSK anticipated that their resuls were "unlikely to go unchallenged". Stewart (1978a) states that the risk estimates by MSK of death from radiation induced cancer were thirteen times higher than those based on atomic bomb survivors and radiotherapy patients. MSK were not disappointed in their expectation that their results would be challenged, and their paper spurred a substantial number of other researchers to reanalyse the Hanford data.

8.2.2 Kneale, Stewart and Mancuso (KSM)

The analysis of the Hanford data by MSK was the subject of an unpublished peer review commissioned by the Energy Research and Development Administration (ERDA). This review was carried out by Dr. Charles Land (see chapter 3) and, despite purporting to show there was no certain evidence of any radiation effects in the Hanford data, found an association between radiation and two types of cancer (myeloma and cancer of the pancreas). In response to such criticisms Mancuso, Stewart and Kneale reanalysed the Hanford data for the 34 year period 1944 to 1977. In addition to using data for an additional five years, the authors made an important adjustment to the data used in their earlier study. Previously they had treated workers who did not wear radiation badges as if they had a zero dose of radiation. In the reanalysis of the Hanford data they excluded any workers who did not wear radiation badges. The data used in the reanalysis is summarised in table 8.4.

Table 8.4 Certified Deaths 1944-1977

	Males		Females	
	Number	Average Dose*	Number	Average Dose*
Cancer deaths	743	2.01	89	0.89
Non-cancer deaths	2999	1.57	202	0.50
Total	3742	1.66	291	0.62

*Average cumulative dose of external radiation measured in rads.

In addition to the 4,033 certified deaths of workers who wore radiation badges detailed in table 8.4, there were 1,497 other deaths which were either not certified, or the worker concerned did not wear a radiation badge, or both. There were also 29,318 Hanford workers who did not die during the period 1944 to 1977. Hence almost 35,000 workers were covered by the reanalysis of the Hanford data, of which some 4,000 were studied in detail.

KSM considered the merits of two alternative statistical techniques for analysing the Hanford data: Comparative Mean Dose (CMD) and SMR. In order to make this comparison it was assumed that the average radiation dose is 1.6 rads (15 mGy) (for the Hanford data it was 1.58 rads (15.8 mGy)), the doubling dose is 30 rads (300 mGy), the normal cancer risk is 2,600 per million, and the results are required to be significant at the 5% level. KSM showed that using the SMR method, a data base of approximately 540,000 man years would be required to produce significant results. This is approximately five times larger than the Hanford data. The CMD method is shown to require data for 145,000 man years to produce significant results, and this is similar in size to the Hanford data. Hence, given the earlier assumptions, the CMD method is capable of producing statistically significant results from the Hanford data whilst the SMR method is not. This vindicates the use of the CMD method by MSK and its continued use by KSM.

MSK controlled separately for five possible causative factors besides radiation, whilst KSM controlled simultaneously for five factors: sex, age at death, date of death, internal radiation and exposure period. KSM applied the Mantel-Haenzel statistical test and concluded that:

(a) controlling for internal radiation strengthens the external radiation effect,

(b) females were more sensitive to radiation induced cancer than males,

(c) those aged between 50 and 60 years were more likely to die of cancer, and

(d) the higher the radiation dose the greater are the cancer deaths.

Thus, for the Hanford data, even when five other factors are simultaneously controlled for, there is still a clear connection between external radiation exposure and cancer. This confirms the results of MSK. KSM then went on to estimate the doubling doses for various categories of cancer, and these are set out in table 8.5.

For all female cancers the doubling dose was only 8.7 rads (87 mGy). These doubling doses are markedly higher than in the MSK study, but there is still a wide gap between these figures and the higher doubling doses based upon atomic bomb survivors and radiotherapy patients. The

Table 8.5 Doubling Doses for Males

Type of Cancer	Doubling Dose in rads*	Doubling Dose in mGy
Bone marrow+	3.6	36
Lung cancer	13.7	137
Cancer of the pancreas, stomach and large intestine	15.6	156
All male cancers	33.7	337

* Average cumulative dose of external radiation measured in rads.
+ Myeloma and myeloid leukaemia.

figures in table 8.5 indicate that it is legally permissible for a worker to receive a doubling dose for myeloma and myeloid leukaemia every year, and a doubling dose for lung cancer every three years. Using these revised risk estimates, KSM calculated that 35 of the Hanford workers died of cancer as a consequence of occupational radiation exposure. This means that roughly 5% of the cancer deaths of Hanford workers were induced by their occupational exposure to radiation. The reanalysis of the Hanford data by KSM confirmed the essential conclusions of the earlier study by MSK. Each study found that the risks of dying from small radiation doses, well within the exposure limits, were much higher than had previously been thought.

8.2.3 Stewart, Kneale and Mancuso (SKM)

Stewart, Kneale and Mancuso (1980) replied to five criticisms of MSK and KSM by the NRPB, and also provided additional tables relating to the KSM analysis. For example, they report that between 1944 and 1974 the annual external doses received by Hanford workers steadily increased over time, as did the number of exceptionally dangerous jobs. After rebutting the criticisms of the NRPB, SKM reaffirm their view that the KSM analysis shows that the risks of radiation induced cancer are ten to twenty times higher than the ICRP estimates. They consider in some detail the criticism that, whilst myeloid leukaemia is thought to be the cancer most likely to be caused by external radiation, MSK and KSM did not find this to be the case. SKM point out that the view that myeloid leukaemia is the most radiosensitive cancer is based upon studies of atomic bomb survivors. They argue that this finding from atomic bomb survivors should not be applied to low occupational doses and that the most radiosensitive cancers to be expected among Hanford workers ought to be myelomas, which is indeed the case.

8.2.4 Kneale, Mancuso and Stewart (KMS)

Using a different technique to those used previously (regression models in life-tables) Kneale, Mancuso and Stewart (1981) reanalysed the KSM data. In the course of this reanalysis of the Hanford data KMS took account of various comments and criticisms that had been made of their earlier studies. The KMS estimates of the dose-response curve found it was concave, and that in the relationship :- Response = (Dose in rads)^x, the maximum likelihood estimate of x was 0.5. KMS also estimated the doubling dose for cancer deaths as 15 rads (150 mGy) and the cancer latency period as 25 years. KMS then attempted to reconcile their estimate of the risks of radiation exposure and those based on atom bomb survivors and radiotherapy patients, where there is a difference of 15 to 20 times.

If a strongly concave dose-response curve fitted to low doses is extrapolated to the higher doses which feature in the atomic bomb and radiotherapy data, the size of the gap between the two curves becomes much smaller. This is shown in figure 8.1. KMS also point out that their estimates do not distinguish between gamma rays and neutrons. Since neutrons are roughly ten times more carcinogenic than gamma rays this may partly explain why the KMS risk estimates are higher than the estimates based upon atomic bomb survivors.

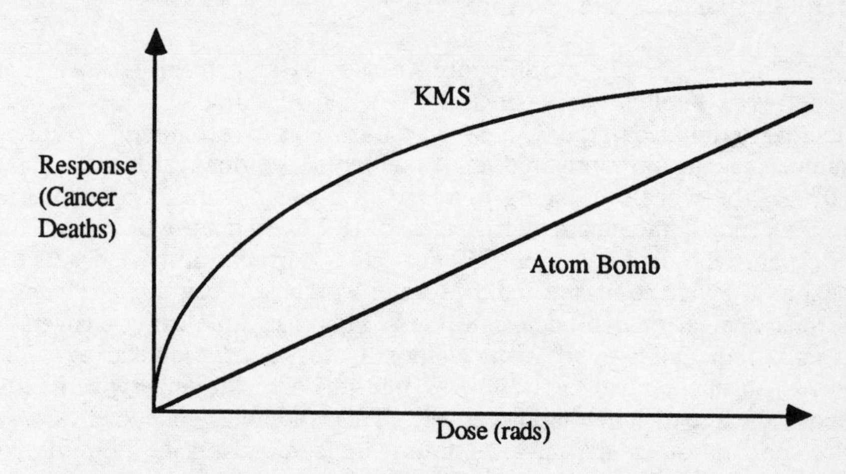

Figure 8.1 Convergence of Concave and Linear Dose-Response Curves

8.2.5 Gofman

Gofman (1979) reanalysed the Hanford data including deaths up to 1977. He omitted any workers who did not survive for at least 15 years beyond the date of hire, all those who had measured internal depositions and all female workers. This left 3,308 deaths, which he divided into those who received a cumulative external dose of more than 10 rads (100 mGy) and the remainder, who received a dose of under 10 rads (100 mGy). After correcting for a difference in average ages at hire of 2.5 years between the two groups, he compared the proportions of deaths from cancer in each of the groups. In the low dose group 25.4% died from cancer, whilst in the high dose group 38.2% died from cancer. The difference is significant at the 5% level. Gofman concluded that his results support those of MSK, KSM, SKM and KMS that low level radiation is many times more dangerous than previous official estimates. Gofman estimated the doubling dose for male cancer deaths to be 43.5 rads (435 mGy) and the increase in cancer deaths resulting from radiation exposure to be 3.5%. The workers in the high dose group had an annual dose rate of approximately 1.1 rads (11 mGy), which is well within the maximum permissible dose of 5 rads (50 mGy) per year. This highlights the result of MSK, KSM, KMS and Gofman that the existing safety standards are not safe. Workers can be receiving only 20% of the maximum permissible dose and yet die of radiation induced cancer.

8.3 Portsmouth Naval Shipyard

The Portsmouth Naval Shipyard in New Hampshire, U.S.A., has undertaken nuclear work since 1959 when the first U.S. operational nuclear submarine, the U.S.S. Nautilus, was overhauled. Nuclear submarines are now repaired and refuelled at the dockyard, with about 20% of the workforce being involved in nuclear work. Najarian et al (1978) traced the death certificates of 1,722 former workers at the dockyard who died between 1959 and 1977. For 592 of these deaths the next of kin were contacted and the former worker was classified as being a former nuclear or non-nuclear worker. This was done using answers to the question 'Did he work with radiation or wear a radiation badge?'. This gave 146 nuclear workers and 379 non-nuclear workers, (with deaths above the age of 80 being discarded). This study was conducted *without* the cooperation of the naval authorities and relied on the help of five reporters from a local newspaper, *The Boston Globe*, to obtain the data.

The study found that for nuclear workers deaths from cancer were 78% higher than expected, whilst for non-nuclear workers the death rate from cancer was only 10% higher than expected. So far as deaths from leukaemia were concerned, the nuclear workers had a death rate 460%

higher than expected, whilst for the non-nuclear workers it was 29% lower than expected. Najarian et al (1978) did not have the radiation records available to them, but as a result of talking to 50 workers they estimated the cumulative doses of external radiation were under 10 rems (100 mSv). This implies that the dose required to double the natural risk of cancer is only about 13 rems (130 mSv), or roughly 25 times lower than previous data would suggest. The Portsmouth study is also reported by Torrey (1978).

The Portsmouth study resulted in a public controversy, and the matter was considered by the U.S. House of Represenatives Subcommittee on Health and the Environment. They commissioned an epidemiological study by the National Institute for Occupational Safety and Health (NIOSH). The resulting study by Rinski et al (1981) received the cooperation of the naval authorities and had access to the medical and dose records of all civilian workers at the dockyard. This is in stark contrast to the initial study by Najarian et al (1978), where access to the data was refused.

Rinski et al studied all white males employed between 1952 and 1977 - a total of 24,545 people. Since all workers in radiation controlled areas were required to wear radiation badges, detailed external dose information was available, and 7,615 workers who had received a recorded lifetime cumulative dose of at least 0.001 rem (0.01 mSv) were identified. The average dose of this group was 2.8 rem (28 mSv) and the median dose was 0.5 rem (5 mSv). The death rates of this group from all causes, all cancers and from leukaemia were compared with the death rates of U.S. white males to compute SMRs. These SMRs are set out in table 8.6.

Table 8.6 SMRs for the 7,615 Exposed Workers

Cause of Death	SMR
All causes	78%
All cancers	92%
Leukaemia	84%

Rinski et al conclude that this indicates there is no evidence for excess cancers. However, the comparison is with all white U.S. males and the SMR for all causes of death of only 78% indicates a considerable 'healthy worker effect'. They point out that the follow-up period was limited, with only 1% of the radiation workers having been followed up for twenty five more years and that an excess cancer rate (relative to the U.S. white male population) may subsequently become apprarent.

Commenting on this study by Rinski et al, Dr. Thomas Najarian, a Boston haematologist, said

> 'I do not agree with the conclusions made in the NIOSH report. They have introduced several fudge factors that tip the numbers to show that nuclear work is safe. If you look carefully at their study, you can only conclude that the whole effort has been a whitewash. It is not fair to compare the nuclear workers against the general population as a whole, because the general population is going to have a higher death rate than healthy workers by a factor of two'.

Torrey (1981). Najarian has also pointed out that the study by Rinski et al excluded 361 workers who were known to have died, but for whom death certificates were unavailable. This reduced the total number of deaths by 7%. In addition, there were 1,012 workers that could not be traced and Rinski et al assumed them *all* to be alive. According to Najarian 'they had no way of knowing whether any of these people were living or dead. Yet for the purpose of this study NIOSH assumed they were all living. And so perhaps there are another 200 deaths that might have elevated the total mortality figures upward', Torrey (1981).

Because of the controversy following the initial Portsmouth study by Najarian et al, the Congress appointed a nine man committee to oversee the NIOSH investigation. The only bio-statistician on this oversight committee was Dr. Irwin Bross, and he has strongly criticised the statistical techniques used in the NIOSH study. Bross has said

> 'I told NIOSH at the outset that the only hope for getting meaningful results in a study like this was to conduct a dose-response analysis of the incidence of cancer among high-dose versus low-dose nuclear workers. This was possible because NIOSH was provided with very good external exposure data by the Navy, with radiation badge records of every man in the study. But they chose instead to ignore the dosage data. They did not use it at all',

Torrey (1981).

Since 56% of the radiation workers received a cumulative dose of under 0.5 rem (5 mSv), their additional exposure to radiation was very small. In view of Bross

> 'obviously, unless we look at the workers who had reasonable exposures of at least one or two rems (10 or 20 mGy), we cannot expect to see any biological effects. Moreover, many of the workers studied had not been exposed long enough to show radiogenic cancers. Not only are the numbers diluted by people who did not get much of a dose, but there are people in there that did receive high doses but could not possibly develop a solid tumour because they had only been at the yard for five years or less. My responsibility as the only bio-statistician on the oversight committee was to warn NIOSH that they were using a flawed statistical technique that could

> only lead to a negative or inconclusive finding. They ignored my recommendation and would not allow the oversight committee to use the Navy dosage data to perform an independent assessment. It sounds incredible, it sounds impossible. But I am convinced that NIOSH had previously decided that they wanted to get a particular result in this study, and they got it',

Torrey (1981).

8.4 Rosyth Naval Dockyard

In 1968 Rosyth Naval Dockyard in Scotland started servicing and refuelling nuclear powered submarines. Blood samples were taken from 197 male workers, starting in 1968 for a ten year period. Dose estimates in rems from film badges were provided by the Admiralty Radiation Records Centre. The radiation sources were nuclear submarine reactors undergoing refit and emitting mixed neutron and gamma radiation, although the exposures were stated to involve almost exclusively gamma radiation. Evans et al (1979) found that, whilst most radiation exposures were below the internationally accepted maximum permissible level for workers of 5 rems (50 mSv) per year, there was a significant increase in chromosome damage. This chromosome damage was found to increase as the dose increased.

8.5 Uranium Miners

Uranium produces the inerts gas, radon, and radon decays to produce various daughter products which become attached to dust particles. When inhaled these dust particles subject the respiratory system to alpha particle radiation. The radiation dose delivered to the lungs of uranium miners from inhaling these dust particles, which may be retained in the lungs, is about twenty times higher than from inhaled radon gas which is exhaled. In chapter 1 the considerable dangers of radiation from internal depositions of radioactive material (i.e. internal radiation) are discussed. As well as being dangerous, internal doses of radiation are much more difficult to measure than are external doses. Radiation badges cannot measure internal doses, and so other methods have to be used. The amounts of radioactive material that a person is excreting can be used to estimate their internal dose. However, this method relies on rather dubious estimates of the relationship between the internal dose and the quantities of radioactive material excreted. This approach is particularly unreliable when low doses of radiation are involved.

Whilst the radiation badges measure the external dose in rads *during* a period of time, the techniques for measuring internal doses quantify the stock of radioactive material within the body *at* some moment in time. Thus internal doses are measured in quantities of radioactive material not in rads, and conversion factors of questionable accuracy are required to produce estimates of the internal dose in rads. Because of these difficulties in measuring internal doses and expressing them in rads, most epidemiological studies have used external dose measurements and, where internal doses have been measured, they are often expressed in quantities of radioactive material and not in rads. So far as the internal dose received by uranium miners is concerned, it is commonly measured in working level months (WLM). This is the number of months for which the person works underground in a uranium mine. The relationship between WLM and rads is problematic since it depends upon the amount of radon gas in the mine and the degree to which dust particles are absorbed into the lungs. Since any number for converting WLMs into rems is likely to be out by a factor of 200 or more, no such factor will be given, Gofman (1983, page 254).

8.5.1 U.S. Uranium Miners

Archer et al (1973) studied 4,146 U.S. uranium miners in Arizona, Colorado, New Mexico and Utah, and found that deaths from respiratory cancer were 500% higher than for a control group of the male population of the four states in which the miners lived. For those with a cumulative exposure of only 120 to 359 WLMs, respiratory cancer deaths were found to be 370% higher than expected. Archer et al (1976) further analysed their data on 4,146 U.S. uranium miners, disaggregating it into 3,366 whites and 780 Indians. Besides extending the follow-up period by six years Archer et al (1976) changed their control group. The Indians were compared with the male non-white population of Arizona and New Mexico, whilst the whites were compared with the male white population of the U.S. They found that deaths from respiratory cancer amongst Indians were 320% higher than expected, whilst for whites the corresponding fiugure was 380%. For the white miners data was available on the length of time worked underground, and the longer a miner had worked underground the more likely he was to die from respiratory cancer. Archer et al (1976) also found some interactions with cigarette smoking, with those smoking the most at highest risk.

Lundin et al (1979) reanalysed the data on the 3,366 white male miners. They found that, for the cancer rate over time, the relative risk model fitted the data better than the absolute risk model. This result supports the view expressed in chapter 2 that the relative risk model is the more realistic. They also found that the median latent period between exposure to radiation and death from lung cancer was roughly ten years. Their findings on the interaction between cigarette smoking, radiation exposure and death from lung cancer were inconclusive. There are two main possibilities for the way in which this interaction can occur: (a) additive and (b) multiplicative. For the additive case, the risk of lung cancer death is equal to the risk from radiation exposure *plus* the risk from cigarette smoking. If the interaction is multiplicative then, in the simplest case, the risk of lung cancer is given by the radiation risk *multiplied* by the smoking risk. All that Lundin et al were able to conclude was that cigarette smokers were at a greater risk of dying from lung cancer than non-smokers, that the number of cigarettes smoked did not appear to affect this extra risk and that more research is necessary to unscramble the interaction between radiation, smoking and lung cancer. The studies of U.S. uranium miners are reviewed in BEIR (1980, pp.318-321).

Samet et al (1984) studied all the cases of lung cancer diagnosed among Navajo men in the U.S. between 1969 and 1981. These 32 cases were compared with a control group of 64 Navajo men who had some other form of cancer. It was found that 23 of those who had lung cancer had a history of uranium mining, while *none* of the controls without lung cancer had been uranium miners. This implies that the risk of lung cancer was over 14 times higher for Navajos who had been uranium miners.

The U.S. Public Health Services have estimated that from 600 to 1,100 out of some 6,000 American uranium miners will die of lung cancer because of radiation exposure, whilst the Australian Atomic Energy Commission found an incidence of leukaemia among white Australian uranium miners that was six times the expected norm, Ryle et al (1980).

8.5.2 Czech Uranium Miners

A group of Czechoslovakian uranium miners who began underground mining in the period 1948 to 1952 were followed up during the years to the end of 1973 by Sevc et al (1976). The exact number of miners studied is not stated but was probably a few thousand. Rates of lung cancer were compared with those of the male population in Czechoslovakia and were found to be 330% higher. The data showed that the longer a miner had worked underground the more likely he was to contract lung cancer. BEIR (1980, page 317) used the figures in the Sevc et al (1976) study to calculate the additional number of cases of lung cancer that would occur if one million people were exposed to one WLM. The BEIR figure is 19 extra cases of lung cancer per million people per WLM.

8.5.3 Canadian Uranium Miners

A study of 15,094 people who worked for more than one month in the Ontario uranium mines is reported in BEIR (1980, pp.321-322). Although exposure to radiation was generally very low, a significant excess risk of lung cancer was observed. For the entire group the lung cancer rate was

80% higher than expected, whilst for those with a cumulative dose of only 1 to 30 WLM's the increase was 64%.

8.5.4 Swedish Iron Miners

Radford (1983) reports the results of a study of 1,400 Swedish iron miners, where the average time spent working underground was twenty years. Their deaths from lung cancer (51 deaths) over the twenty five year period 1951 to 1976 were recorded. The radiation levels from radon daughters in Swedish mines were similar to the current standard for U.S. mines. It was found that after a miner had been working underground for over ten years there was an increase in the risk of dying from lung cancer of between three and four times, as compared with Swedish national rates. This increase was statistically significant at the 1% level. The increase in the death rate appeared about twenty years after miners began work underground and continued for at least another thirty years, implying that the latent period for lung cancer can be over fifty years.

These results for Swedish iron miners suggest that, at a cumulative dose of twenty seven WLM's, the risk of dying from lung cancer is doubled. This dose of alpha radiation is only about five times that from radon daughters inside the home. The study also suggests that current U.S. standards for miners permit a three to four fold increase in lung cancer deaths. Radford also looked for synergy between cigarette smoking and radiation exposure in causing lung cancer deaths, but none was found.

8.5.5 British Iron Miners

Iron ore has been mined in West Cumbria since Roman times. In 1969 a survey found high concentrations of radon gas in the air of three of the four mines in the Egremont area, with all the measurements being above the ICRP maximum permissible level. This suggests that the dose of radiation to the lungs of the miners may have led to deaths from lung cancer, and Boyd et al (1970) studied the death rate from lung cancer among iron ore miners. They examined the 5,811 death certificates of all the males who died between 1948 and 1967 whilst resident in Ennerdale Rural District and Whitehaven Metropolitan Borough. This revealed that 36 underground iron ore miners had died from lung cancer whilst the number of such deaths expected using either national rates or local rates (excluding iron ore miners) was only 21. Thus the lung cancer mortality rate was 71% higher than expected, and this excess was statistically significant at the 0.1% level.

8.6 Uranium Processing

There do not appear to have been studies of the health effects of radiation on workers employed in processing uranium ore. However, there has been a small study in Brazil, which is of relevance. A mill, which processes 150 tons per month of monazite and employs 420 people, is located in a densely populated borough of Sao Paulo, Brazil. Monazite sand contains up to 6% of thorium oxide and 0.3% of uranium oxide (U_3O_8). A study by Costa-Ribeiro et al (1975) found a statistically significant increase in the number of chromosome aberrations amongst workers with the highest exposure at the plant.

8.7 Rocky Flats Plutonium Workers

Brandom et al (1978) studied chromosome aberrations amongst 343 workers at the Rocky Flats Nuclear Weapons Facility in Colorado who had been exposed to radiation from plutonium in the course of their employment. Studies of the effects of plutonium emissions from Rocky Flats on the cancer rates of the surrounding population will be discussed in chapter 9. Brandom et al also studied 80 Colorado uranium miners who had been exposed to radiation. They found that, for both plutonium workers and uranium miners, the number of complex chromosome aberrations in the blood per unit of radiation *declined* as the radiation dose increased. This implies that the dose-response curve is concave. Using U.S. safety standards Brandom et al calculated the number of complex chromosome aberrations that would be expected to result from the maximum permissible lifetime exposure to plutonium and uranium mining. They found that the plutonium worker would have roughly twenty times more complex chromosome aberrations than the uranium miner. This suggests there is an inconsistency in exposure standards between plutonium workers and uranium miners, with the exposure standard for plutonium workers apparently twenty times more lax.

8.8 Windscale (Sellafield)

At Windscale (renamed Sellafield in 1981) in Cumbria are three nuclear reactors (of which two have been shut down), two reprocessing plants, a plutonium finishing plant and a mixed uranium plutonium oxide plant. Windscale (Sellafield) has been involved in reprocessing spent nuclear fuel since 1952. The only study of the effects of radiation upon the health of the Windscale (Sellafield) workforce, currently about 10,000, is the NRPB study referred to in chapter 3. This study is controversial. The NRPB claims it shows there are no health risks, while others argue that the study was badly conducted and that if some corrections are made there does appear to be an association between radiation dose and myelomas and RES neoplasms, Scott et al (1980, page 153).

In 1977 a public inquiry was held into plans to build a thermal oxide reprocessing plant (THORP) at Windscale (Sellafield). During the course of this inquiry various statements were made concerning the level of occupational exposure at Windscale (Sellafield) and the possible health consequences for the workforce. Ellis (1978) reports that between 1971 and 1975 30% of the fuel reprocessing workers at Windscale (Sellafield) were receiving radiation doses above the maximum permitted level of 5 rems (50 mSv) per year. In evidence to the Windscale (Sellafield) Inquiry Ellis said of the radiation exposure of the workforce that 'these higher levels are probably the highest levels of exposure reported in any industry involving radiation in the U.K. or elsewhere in the world', Stott et al (1980, page 156). A witness on behalf of BNFL (who operate the Windscale (Sellafield) plant) stated in evidence to the Windscale (Sellafield) Inquiry that, using the ICRP risk factors, eight workers at the plant are estimated to have been killed by radiation induced cancer. If larger risk factors are used, as has been suggested by many of the studies reported in this book, the estimated number of radiation induced deaths is greater than eight.

In 1957 there was a major fire in a reactor at Windscale (Sellafield), and this serious nuclear accident will be discussed in chapter 9. Whilst dealing with this fire, some workers were exposed to high external doses of radiation and also to significant air concentrations of radionuclides, leading to internal doses. Some of these workers were found to have received thyroid doses of 9.5 rems (95 mSv), Crick et al (1982, page 20).

On various occasions statements have been made to the effect that no workers have been killed and/or injured by their exposure to radiation. For example, Dr. Hans Blix, Director General of the IAEA, is quoted in Atom (1983a) as saying that 291 nuclear reactors have accumulated an operating record of 2,800 reactor years without a single radiation-induced fatality. Mr. John Dunster, director of the NRPB, has said there is no direct evidence that occupational exposure to radiation at the levels achieved for workers in British industry has caused any ill-health or death, Financial Times (1983a). These statements, and many others like them, are incorrect.

Much evidence to the contrary is available throughout this book e.g. section 8.13 of this chapter. Some examples of where liability for deaths of workers at Windscale (Sellafield) has been admitted, will now be presented. The first payment under a compensation agreement between BNFL and the Associated Union of Engineering Workers (AUEW) was made at the end of 1983. BNFL paid £21,654 in compensation to the dependents of a former Windscale (Sellafield) worker who was exposed to radiation and who died of leukaemia in 1965, Atom (1984c). In 1985

BNFL paid £120,000 under the compensation agreement to the family of a 45 year old man who died from cancer in 1976. During his 23 years employment at Windscale (Sellafield) he had received a total dose of 60 rems (600 mSv) i.e. only 52% of the exposure limit, Rose at al (1985). In 1985 BNFL agreed to pay £31,000 compensation in respect of a BNFL worker who died from stomach cancer in 1978, Atom (1985e).

During the period 1971-83 BNFL have also made compensation payments to five other Windscale (Sellafield) workers in respect of radiation induced deaths, Atom (1983b). There were two deaths from myeloid leukaemia and one death each from brain cancer, cancer of the pancreas and multiple myelomatosis. In October, 1984 an inquest jury decided that the death of Mr. Isaac McAllister of Whitehaven in June 1983 from bone cancer was caused by his job. Mr. McAllister had worked at Windscale (Sellafield) for thirty years, and had received a measured external dose of 86 rems (860 mSv) during his employment, Guardian (1984b). In 1986 the dependents of three workers who died from cancer received compensation from BNFL, in one case amounting to £57,000.

8.9 Sizewell

There have recently been five leukaemia cases among workers at the existing Sizewell A nuclear power plant, and all the men concerned were employed inside the reactor area, where the radiation exposure is highest. Details of these five cases are set out in table 8.7, which is based upon Tucker (1982). This shows that all these cases occurred during a two and a half year period.

Table 8.7 Leukaemias Among Sizewell Workers

Name	Outcome	Date
Ken Thompson	Death	August 1980
Robert Biddle	Diagnosis	January 1981
George Marjoram	Death	November 1981
Michael Hope	Death	March 1982
Tony Adams	Death	February 1983

The number of workers at Sizewell is roughly 400 to 500 people and, in the view of Dr. Alice Stewart, it is unlikely that all these cases have been caused by chance, Chorlton (1983). The local TGWU organizer, Mr. Len Chapman, has said that 'the authorities are not going to admit the connection between radiation work and leukaemia unless they are forced to', Chorlton (1983).

8.10 UKAEA Workers

A study by the London School of Hygiene and Tropical Medicine has examined mortality amongst 39,546 people employed by the UKAEA at Harwell, Dounreay, Winfrith, Culham and London between 1st January 1946 and 31st December 1979, Beral (1985). This is one of the largest studies of occupational exposure to radiation. Data from badges was available on the external radiation dose received during employment by the UKAEA, and this averaged 3.24 rems (32.4 mSv) for the 19,164 exposed workers. They found that cancer of the prostate was significantly related to the cumulative radiation dose received by the worker. The death rate from prostate cancer was also compared with the national rate. Due to the 'healthy worker' effect (see chapter 2) the UKAEA employees may be expected to have a relatively low death rate from prostate cancer. But the opposite is the case and, for those workers who were exposed to tritium and who received a cumulative dose of over 5 rems (50 mSv), the death rate from prostate cancer was 13 times higher than the national rate. This increase was statistically significant at the 0.1% level.

Beral et al also fitted dose-response curves to their data for deaths from leukaemia and all cancers. In both cases their estimated risks per rem were three times higher than the ICRP estimates, although the possible errors were very large. Thus, they were unable to exclude the possibility that the ICRP risk figures are fifteen times too low, or that exposure to radiation is beneficial! The problem is that this study did not have sufficient data to detect an increase in radiation related cancers unless the ICRP risk factors understate the true risks by a factor of about twenty or more - see chapter 4 for a discussion of sample size.

8.11 Radium Dial Painters

Studies of persons who have ingested large amounts of radium (mainly radium dial painters and scientists) indicate that the dose-response curve for cancer rises to a peak and then falls slightly for higher doses, Morgan (1973). Table 8.8 from Morgan (1973) shows that, as the average bone dose increases, biological changes become more common and more serious.

In 1915 a Dr. von Hoffman set up the Radium Luminous Materials Company in New Jersey employing hundreds of girls, some as young as 14, to paint the dials of watches, light bulbs and crucifixes with a mixture of radium and zinc sulphide, Morgan (1984). Dr. Sabia von Sochocky had invented the formula for the radium paint in 1913. Later, when concern was mounting over the health risks, he said there was no possible association between his paint and the deaths of radium dial painters. He

Table 8.8 Long Term Effects of Radium in Man

Average Bone Dose (rem per year)	Minimal	Biological Changes % Mild	Moderate	Advanced
0.3 to 9	8%	0	0	0
9 to 30	13%	0	4%	0
30 to 90	15%	3%	6%	6%
90 to 300	25%	25%	16%	22%

subsequently died of cancer at the age of forty-six and had a level of radium in his body higher than that found in many of the dial painters, Barlett et al (1985, pp.307-309).

Polednak et al (1978) have studied 634 women who worked in the U.S. radium dial painting industry between 1915 and 1929. Their mortality rates were compared with those for U.S. white females. For bone cancer the mortality rate was over 80 times higher than expected. Disaggregating the data into year of first exposure, the increased rates for bone cancer for the period 1915 to 1919 were 230 times; for 1920 to 1924; 150 times and for 1925 to 1929; 9 times. This indicates a dramatic drop in the excess mortality rate for women starting work after 1924. The main determinant of radium dose was the woman's work habits, in particular whether and how often they licked their paint brushes. Accuracy was achieved when the tips of the camel hair brushes used to apply the paint were finely pointed, and the girls soon developed the habit of rolling the brushes between their lips. Some of the girls even painted their teeth with the radium mixture before an important 'date' so that they would glow romantically in the dark, Morgan (1984).

The decline in excess bone cancer rates corresponds with a change in work regulations as in 1925 the tipping or pointing of the brush between the lips was prohibited. Najarian (1978) reports that for as long as five years after the discovery that radium dial painters were being killed by their occupational exposure to radium, a New Jersey company that employed the women continued to deny the deaths had anything to do with employment at the factory.

In addition to the bone cancer induced by internal radiation from swallowing radium, external radiation of the body by the paint pots led to an increase in breast cancer. This effect was not identified until 1980 because it tended to be dominated by the highly significant increase in bone cancer, Pochin (1938, page 131).

A study of 1,235 women who worked in the radium dial painting industry in the U.S.A. before 1930 was carried out by Stehney et al (1978). They studied the deaths of these women up to 1976, so that between 45 and 60 years of observation after initial exposure were possible. They compared the death rates of these women with those for U.S. white females and found their death rate from *all* causes was 15% higher. This increase in the death rate was significant at the 0.5% level.

8.12 Radiologists

In the course of their occupations, the early radiologists received exposure to X-rays. A study covering the period 1935 to 1958 found that, for radiologists aged between 50 and 64 years, the death rate from leukaemia was over seven times higher than that of similar medical workers not exposed to X-rays, Morgan (1973). It was also found that the early radiologists' life expectancy was shortened by 4.8 years in the period of 1935-1944, four years in the period of 1945-54 and 2.9 years in the period of 1955-58, Morgan (1973). Later studies indicate that from about 1960, when radiologists received lower doses of X-rays (mainly due to diagnostic X-rays being delivered by the X-ray technologist and not the radiologist) there has been no detectable life shortening of radiologists, Morgan (1973).

8.13 Fatal Accidents

Some examples were given in chapter 4 and in section 8.8 of this chapter of statements to the effect that radiation has not killed any workers. Such statements are manifest nonsense. The details of the deaths of ten workers who have been killed by nuclear accidents involving massive radiation doses are given below, UNSCEAR (1982, pp. 413-415). The first eight deaths occurred in the nuclear industry whilst the last two deaths involved non-nuclear workers. For each accident the date, location, whole body dose, period to death and details of the accident are given.

(a) 1945 / Los Alamos, USA / 300 rads (3,000 mGy) / Death after 25 days. A worker was stacking tungsten carbide bricks around a plutonium core and accidentally made the system critical. He remained to dismantle the assembly.

(b) 1946 / Los Alamos, USA / 1,200 rads (12,000 mGy) / Death after 9 days. A worker was demonstrating the creation of a critical assembly when a beryllium shell fell into the assembly.

(c) 1958 / Los Alamos, USA (plutonium recovery plant) / 4,500 rads (45,000 mGy) / Death after 35 hours. Excess plutonium was washed into a large vessel when the operator started a stirrer.

(d) 1958 /Vinca, Yugoslavia (experimental reactor) / 436 rads (4,360 mGy) / Death after 4 weeks. The reactor became uncontrolled when the amount of heavy water moderator was accidentally increased.

(e) 1961 / Idaho, USA (reactor) / 30-35,000 rads (300-350,000 mGy) / 3 immediate deaths. Explosion, probably caused by the excessive withdrawal of the central control rod. One of these men was impaled on the ceiling by the force of the accident. The bodies of these three men had to be dismembered because parts of their bodies were so highly radioactive that they had to be buried in lead lined coffins, Bertell (1985, page 206).

(f) 1964 / Wood River Junction, USA (uranium recovery plant) / 1,200 - 4,600 rads (12,000-46,000 mGy) / Death after 46 hours. A technician mistakenly poured a highly radoactive liquid into a tank.

(g) 1960 / USSR / 14,800 rads (148,000 mGy) / Death. A demented worker placed a caesium-137 source in his trouser pocket.

(h) 1975 / Italy / 1,000 rads (10,000 mGy) / Death. A worker was exposed to cobalt-60 radiation at an agricultural installation.

These deaths are clearly attributable to radiation because the workers concerned exhibited the signs of radiation sickness and died shortly after exposure. Very many more deaths have resulted from cancers occurring years after the radiation exposure, but the connection is less obvious.

8.14 Increased Risk of Cancer from Occupational Exposure

Bonnell et al (1978) have compared the possible cancer risks from occupational exposure to radiation, with the natural risks of cancer. They used the ICRP risk factors to calculate the cancer risks from occupational exposure to radiation. For a worker receiving the maximum permissible radiation dose of 5 rems (50 mSv) per year between the ages of 20 and 65, the ICRP risk factors can be used to calculate the resulting increase in deaths from cancer and leukaemia. The natural rates of cancer deaths and leukaemia used were those for males in England and Wales for 1969-73, adjusted for age. The increase in the cancer and leukaemia rates resulting from maximum permissible occupational exposure are expressed as a percentage of the natural rates in Table 8.9.

Table 8.9 Percentage Increase in Deaths when there is Maximum Permissible Radiation Exposure

Age	Increase in Cancer Deaths	Increase in Leukaemia
35	19%	217%
45	23%	250%
55	11%	135%

As argued throughout this book, the true cancer risks from radiation may be substantially greater than indicated by the ICRP figures used in the calculation of table 8.9. Hence the percentage increases in this table may be much too small. Bonnell et al point out that the measured radiation doses to most workers are less than the maximum permissible dose used in calculating the figures in table 8.9. But the conclusion still stands, that occupational exposure can lead to a very large increase in leukaemia rates and a substantial increase in cancer deaths.

In the U.S., the EPA standard is that no worker should be exposed to more than 0.5 rem (5 mSv) per year. Lapp (1979, page 72), presents a summary of the official returns made to the Nuclear Regulatory Commission (NRC) in 1976, and part of these data are given in table 8.10. The NRC calculated the average dose per worker to be 0.36 rem (3.6 mSv). These data illustrate the general point that whilst the average dose per worker may be low, particularly if many workers who received a zero dose are included, there may well be workers who received a considerably higher dose. Table 8.10 shows that, in their official returns to the NRC, the operators admitted to exposing 386 workers to a dose in excess of the ICRP limit of 5 rems (50 mSv) per year. Hence statements that exposures are well within the standards, with the average dose being only 7% of the limit, should be considered in the light of the above evidence.

Table 8.10 Annual Whole Body Doses in the U.S. in 1976 by Occupational Group - Numbers of Workers

Occupational Group	Annual Dose in Rems							
	5	6	7	8	9	10	11	12
Nuclear reactors	188	70	26	11	5	1	0	0
Industrial radiography	15	10	3	2	0	2	0	3
Fuel processing	17	0	0	0	0	0	0	0
Manufacturing & distribution	16	10	5	2	0	0	0	0
Total	236	90	34	15	5	3	0	3

There are some occupational groups where the average dose is well above 0.36 rem (3.6 mSv) per year, e.g. mechanical maintenance workers at Canadian reactors - 1.2 rads (12 mGy) in 1979, BNFL workers at Windscale (Sellafield) - 0.8 rads (8 mGy) in 1978, and U.K. gas luminizers - 1.6 rads (16 mGy) in 1976, UNSCEAR (1982, pp.401-402 and page 411). The average exposures of workers at the U.S. reprocessing plant in West Valley, New York were 1968 - 2.74 rems (27.4 mSv), 1969 - 3.81 rems (38.1 mSv), 1970 - 67.6 mSv) and 1971 - 7.15 rems (71.5 mSv). With average worker exposures increasing each year, the plant was closed in 1972, Bertell (1985, pages 268 and 417).

8.15 Summary

The studies considered in this chapter indicate that there may be significant health risks associated with being occupationally exposed to radiation. The studies of the radium dial painters found a very large increase in the cancer rate following exposure to considerable doses of radiation. This led to a subsequent tightening of the safety procedures for radium dial painters. The studies of uranium miners in the U.S.A., Czechoslovakia and Canada indicate an increase in lung cancer for those engaged in this occupation.

The Rosyth and Rocky Flats studies are concerned with chromosome damage not cancer. Whilst the sample sizes were not very large, each study indicates an increase in chromosome aberrations amongst workers exposed to low levels of radiation. The initial Portsmouth study lacked a solid data base due to the refusal of the naval authorities to supply the data in their possession, but it suggests that low levels of occupational exposure result in an increase in cancer. The subsequent NIOSH study has not resolved whether there was an increased cancer risk.

The Hanford study was concerned with the effects of small doses of radiation which were well within the safety limits. Whilst the findings of the Hanford study remain controversial, it is widely accepted that it does demonstrate an increase in certain types of cancer amongst the workers exposed to radiation. What remains controversial is the extent to which radiation exposure has caused an elevation in cancer rates.

9 Public exposure to radiation from nuclear establishments

9.1 Introduction

People who are occupationally exposed to radiation are usually aware that they are working with radioactivity and may have some idea of the risks they are running. They also have some degree of choice over whether or not they are occupationally exposed to radiation. Those who are exposed to radiation whilst undergoing medical procedures may realize they are being irradiated and that this is a risky procedure. They may also be able to refuse to subject themselves to such procedures, e.g. being X-rayed. However, members of the public who are exposed to radiation from nuclear establishments, e.g. nuclear power stations, nuclear bomb factories, reprocessing plants, etc., are usually unaware that such exposure is taking place. Since they are in ignorance of the fact that they are receiving an involuntary dose of radiation they are unable to choose whether or not to continue absorbing this dose.

The public are involuntarily exposed to radiation from various other sources besides nuclear establishments, such as power stations. In chapter 7 the adverse effects upon the population of nuclear bomb tests and the highly adverse consequences for the Japanese population of the dropping of bombs on Nagasaki and Hiroshima have been considered. In chapter 10 the effects of exposure to background radiation, plutonium in water, uranium tailings and smoking will be considered. In this chapter the health consequences for the population living near nuclear reactors, nuclear bomb factories and reprocessing plants will be considered. Since these studies are concerned with the health effects on the public and not on employees, the basic data are not secret. Furthermore, the number of people involved can be substantial, so increasing the chances of detecting relatively small changes in cancer or infant mortality rates. But the task of collecting and analysing the data can be formidable, and there is a lack of data on individual doses. These, therefore, have to be estimated. There is also the problem of controlling for all other factors, e.g. smoking habits. The Rocky Flats study appears to have overcome these difficulties successfully, although the other studies considered in this chapter are somewhat less refined.

9.2 Rocky Flats

In Colorado the Rocky Flats Nuclear Weapons Facility has manufactured plutonium triggers for nuclear warheads and repaired and replaced defective bomb and warhead components since 1953. Johnson et al (1976), Johnson (1979) and Anon (1977), found that about two miles downwind of Rocky Flats the levels of plutonium in the soil were 380 times the level that could be accounted for by fallout. Johnson (1979 and 1981) reports further that in some places the levels of plutonium in the soil were 3,390 times the level due to fallout and that plutonium 239 was the predominant radioactive material, but plutonium 238, 240 and 241 and americium 241 and caesium 137 were also present. Figure 9.1 shows that the plutonium contamination is centred on the Rocky Flats plant and that it reaches well into the city of Denver. In Shutdown (1979, page 50) Dr. John Gofman testifies that the AEC estimated that only a milligram of plutonium had been released into the surrounding areas by the Rocky Flats plant. However, Dr. Edward Martell found that in fact about half a pound of plutonium had escaped. This estimate was subsequently confirmed by the AEC. Not only was the AEC estimate of plutonium releases an underestimate by two hundred thousand times, but the amount of plutonium actually released was potentially capable of killing 200 million people.

Further studies by Johnson (1979, 1980 and 1981) analysed cancer rates in 1969, 1970 and 1971 amongst the one million people living in the Denver area downwind of Rocky Flats (the centre of Denver is 16 miles downwind of Rocky Flats). This epidemiological study found that cancer rates were increased up to 24 miles downwind of the plant, with those living closest to the plant being most at risk. A summary of Johnson's results appear in table 9.1.

All of the results in table 9.1 (except the 4% increase) are statistically highly significant. The study found significant increases in a wide variety of cancers in the population living up to 24 miles downwind of Rocky Flats, and a selection of the higher increases is summarized in table 9.2.

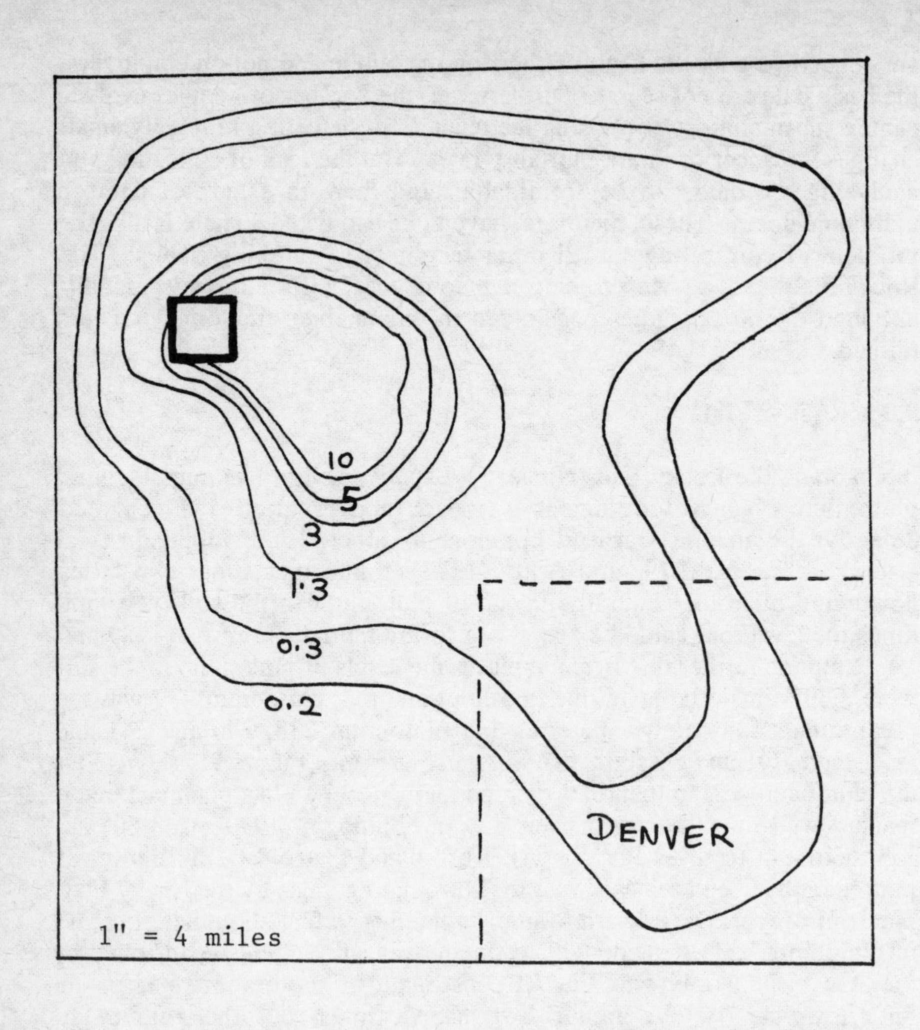

Figure 9.1 Contours of Plutonium Dust Round the Rocky Flats Plant, in millicuries

Table 9.1 Increased Cancer Rates by Distance from Rocky Flats

	Increase in Cancer Rates		
Distance Downwind of Rocky Flats	Male	Female	Total
0 to 13 miles	24%	10%	16%
13 to 18 miles	15%	5%	10%
18 to 24 miles	8%	4%	6%

Table 9.2 Increases in Various Types of Cancer

Type of Cancer	Increase in Men
Lung & Bronchus	34%
Leukaemia	40%
Lymphoma & Myeloma	43%
Colon & Rectum	43%
Testes	140%

Type of Cancer	Increase in Women
Ovaries	24%
Colon & Rectum	30%

Over a three year period, 501 people out of a population of 595,226 living in the study area up to 24 miles downwind were estimated to have developed cancer as a result of the Rocky Flats plant. The details of the calculation of these extra cancers appear in table 9.3.

Table 9.3 Extra Cancers Caused by Rocky Flats

	Male	Female	Total
Cancers observed	2808	2939	5747
Cancers expected	2456	2790	5246
Extra cancers	352	149	501

The increase in cancer rates in the Denver area may be due to the presence of plutonium released into the atmosphere by Rocky Flats. This interpretation is supported by the way the higher cancer rates are associated with levels of plutonium in the soil, and the distance from Rocky Flats. The higher levels of plutonium downwind of Rocky Flats may be due to the normal functioning of the Rocky Flats plant and/or accidents at Rocky Flats. Johnson et al (1976), Johnson (1979, 1980 and 1981) and Anon (1977) report that there have been two serious fires at Rocky Flats.

The Rocky Flats study has been criticised by Reissland et al (1980) of the NRPB on the grounds that (a) it did not make allowance for other factors besides age, sex and race which may have caused variations in cancer rates, (b) it used the least contaminated suburban area of Denver as the control, (c) there was uncertainty about the plutonium levels in the soil, and (d) the statistical methods employed were suspect. These criticisms have been answered by Johnson (1981).

A new study of rates in the Denver area between 1969 and 1979 has been carried out by Chinn (1981) and this study overcomes the four criticisms that have been made of Johnson's study. Chinn used a different statistical technique from Johnson; multiple regression analysis. This technique readily permits the effects of other factors to be allowed for, and Chinn considered 13 socioeconomic variables, 3 air pollution variables, 5 land use variables, 5 occupational variables, 13 population mobility variables and 2 variables on other possible environmental sources of radiation. Due to the alleged uncertainty about the plutonium levels in the soil, Chinn used 9 different measures of the Rocky Flats dose (one of which was plutonium levels in the soil). Finally, the control group used by Chinn was the cancer rate for the entire Denver area, including the contaminated area, so tending to *understate* any risk.

Chinn's results are in broad agreement with those of Johnson. The most important variable explaining variations in cancer rates by census tract is the Rocky Flats dose. The inclusion of the other variables, if anything, strengthens the estimated influence of the Rocky Flats dose on cancer rates. The best measure of the Rocky Flats dose was the extent to which the census tract was downwind of Rocky Flats. This variable was measured by the cosine of the angle between the census tract and the average wind direction from Rocky Flats over the period 1953 to 1970.

Chinn also used the same statistical technique as Johnson - contingency table analysis - but allowing for differences in income and air pollution as well as age, sex and race. A comparison of the various estimates of the increase in cancer rates among the 181,116 people living immediately downwind of Rocky Flats, that is within about 13 miles, is set out in table 9.4.

Table 9.4 Increase in Cancer Rates Relative to the Average for the Denver Area

	Male	Female	Total
Chinn: Multiple Regression Analysis	+ 18%	+ 7%	+ 16%
Chinn: Contingency Table Analysis	+ 16%	+ 5%	+ 10%
Johnson: Relative to his control area	+ 24%	+ 10%	+ 16%

Table 9.4 reveals that the various authors and techniques are in broad agreement about the level of increase in cancer rates in the most contaminated area. Hence Johnson's earlier results are confirmed by Chinn (1981), in a study that has overcome the objections levelled against the earlier study. This means that Johnson's initial claim that Rocky Flats is killing hundreds of people living downwind of the plant appears to be substantiated.

9.3 Windscale (Sellafield)

There has been no major study of the health effects due to the operation of the various nuclear facilities at Windscale (Sellafield) in Cumbria on the surrounding population. (The effects upon the workforce were discussed in Chapter 8.) However, there is evidence on the levels of radioactive material which Windscale (Sellafield) has been emitting into the environment and of the resulting increase in the levels of radiation in surrounding areas. There is some data on the adverse effects on the health of the local population, e.g. the Black report, and it is also possible to estimate the likely health effects of the measured environmental emissions using risk factors estimated in other studies. Figure 9.2 shows Cumbria and the location of Windscale (Sellafield), Drigg and Seascale.

9.3.1 Environmental Effects of Windscale (Sellafield)

The nuclear fuel reprocessing plant at Windscale (Sellafield) is discharging water contaminated with radioactive material (plutonium and americium). This is a clear dilute and disperse solution to the problem of disposing of low level waste. However, although the radioactivity is being released into the environment, it is not dispersing as expected. Day et al (1981) have taken samples of sediment from the Ravenglass Estuary adjacent to Windscale (Sellafield). Both the americium and plutonium seem to be largely retained in sediments relatively close to Windscale (Sellafield) and not dispersed throughout the Irish Sea as expected. Deposits of the highly dangerous americium 241, which emits alpha particles, are building up from both direct deposition and from the radioactive decay of sedimented plutonium 241. It is estimated there are 18,000 curies of americium 241 and 382,000 curies of the beta emitter, plutonium 241. Lean et al (1984) have estimated that a quarter of a ton of plutonium now lies at the bottom of the Irish Sea. The maximum inhalation allowable in occupational exposure is a millionth of a curie per year. Not only may these radioactive deposits contaminate the food chain, but they may move shoreward and be resuspended as particles in the atmosphere.

The Secretary of State has said it would take a cataclysm to remobilise the sediment in the Ravenglass estuary. And yet Stott et al (1980, page 163) have compared the levels of plutonium and americium in the air at Ravenglass during a four week period in August/September 1977 with the levels in Oxfordshire for the same period. The levels of plutonium 239 and 240 at Ravenglass were 10 times higher, whilst for americium 241 they were about 50 times higher. The NRPB stated that they did not think these figures warranted any further investigation! From the Solway Firth to Barrow the contamination of sea spray with plutonium and americium is

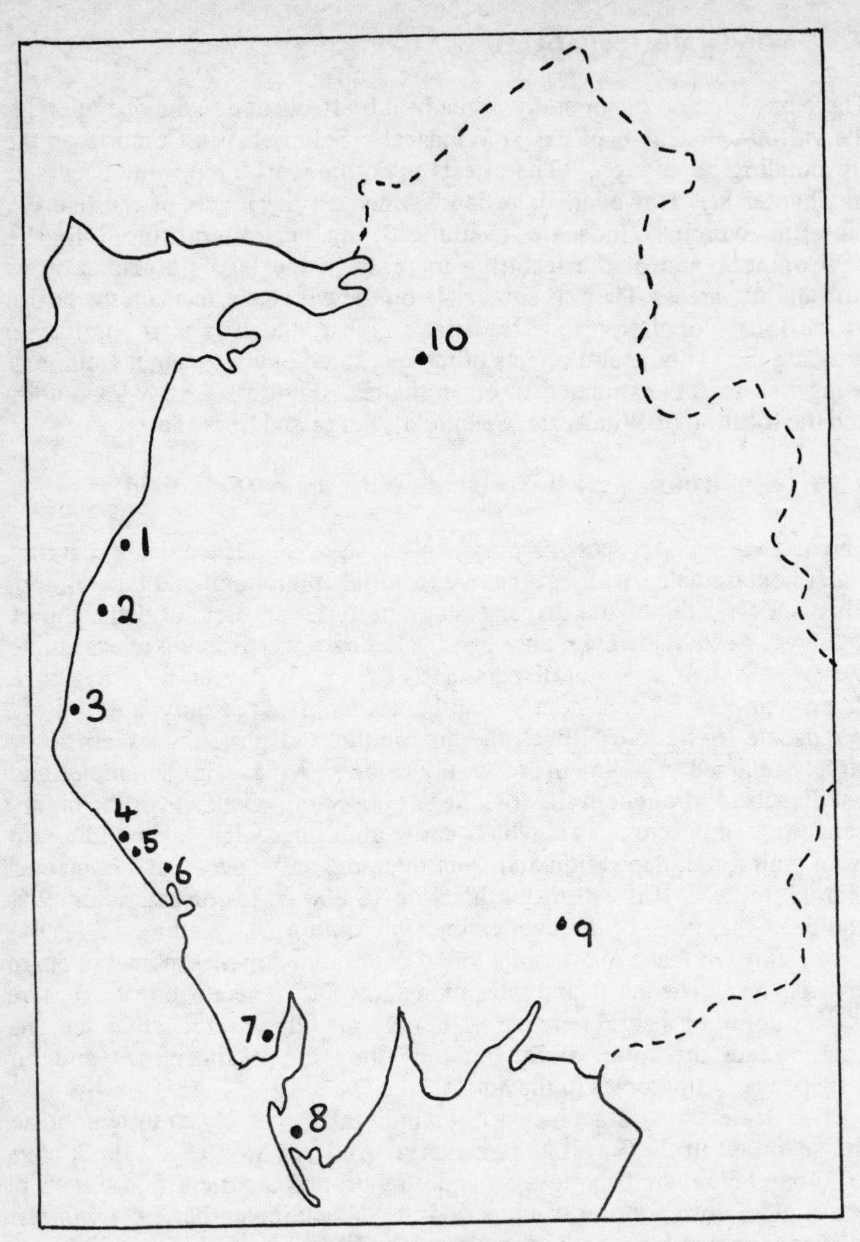

Figure 9.2 Cumbria

1 = Maryport, 2 = Workington, 3 = Whitehaven, 4 = Windscale, 5 = Seascale, 6 = Drigg, 7 = Millom, 8 = Barrow in Furness, 9 = Kendal, 10 = Carlisle

readily detectable. Whilst the caesium-137 tends to remain in solution in the sea water, the plutonium -239 and 240 and americium -241 that contaminates the sea spray comes from suspended sediments in the surf zone. A study by Eakins et al (1982) has found that the concentrations of plutonium and americium in the sea spray may be very much greater than in the sea water. Table 9.5 shows that enrichment of several thousand times may occur.

Table 9.5 Mean Enrichment Factors for Sea Spray

Element	Unfiltered Sea Water	Filtered Sea Water
Plutonium 239 and 240	15	330
Americium 241	15	3,700

If the plutonium and americium discharged by Windscale (Sellafield) is resuspended and blown ashore, there should be evidence of this in the coastal regions around the plant. Eakins et al (1985) studied the sediments in Harnsey Moss Tarn, which is 2.5 miles up the coast from the plant and 100 yards from the sea. They concluded that the sediments of this lake contained plutonium and americium, and that the predominant source of this radioactive material was sea discharges by Windscale (Sellafield).

Taylor (1982, pages 33, 40 and 41) reports the results of various studies of the levels of plutonium in the air, soil and coastal waters of Cumbria. The levels of plutonium 239 in seawater off the Cumbrian coast were found to be 2,000 times higher than would be expected due to fallout. Levels of plutonium 239 and 240 in the soil around the Windscale (Sellafield) plant were found to be over 7 times higher than would be expected due to fallout, Cambray et al (1982) and Peirson et al (1982). The concentration of plutonium in the soil decreased in roughly a north easterly direction from the plant. Levels of plutonium suspended in the air along the Cumbrian coast were about 10 times higher than would be expected due to fallout.

In 1984 Fry et al (1985) conducted a small-scale comparative study of the radioactive materials present in house dust in 20 houses in West Cumbria and 5 houses in Oxfordshire and Berkshire. They found a very marked difference between the two areas. Table 9.5 shows the number of times the amount of radioactivity in West Cumbrian houses exceeded that in Oxfordshire/Berkshire.

A huge disparity exists between the measured sea discharges of radioactive material from Windscale (Sellafield) and those from other UK nuclear establishments. Table 9.7 shows that in 1982 Windscale (Sellafield) accounted for 98.8% of the UK total, Atom (1984a).

Table 9.6 Relative Quantities of Radioactive Materials in West Cumbrian and Oxfordshire/Berkshire Houses

Caesium - 137	17 times
Plutonium - 238	250 times
Plutonium - 238 & 239	55 times
Americium - 241	1,375 times

Table 9.7 Discharges to the Sea in 1982

Nuclear Establishment	Curies
Windscale (Sellafield)	83,236
Dounreay	446
Hunterston A & B	85
Sizewell A	76
Bradwell	74
Berkeley	74
Dungeness A	56
Oldbury	49
Hinkley Point A & B	41
Chapelcross	32
Wylfa	20
Winfrith	5

Liquid and atmospheric discharges of radioactive material from Windscale (Sellafield) are monitored by BNFL. Thus BNFL monitor themselves! They report the results to the authorizing government department for publication in annual reports, Black (1984, page 53). The consequential environmental effects of the discharges are also monitored by BNFL, subject to the requirements of the Department of the Environment and the Ministry of Agriculture, Fisheries and Food (MAFF). MAFF also undertake marine monitoring, whilst various other bodies monitor the environment from time to time. The responsible government department will presumably compare the radioactive discharges reported by BNFL with the authorized limits. But for atmospheric discharges this check is impossible because there are *no* quantitative limits on atmospheric discharges from Windscale (Sellafield), Black (1984, page 51).

It is revealing to note that the monitored levels of radioactive material released by the French reprocessing plant at Cap de la Hague are massively less than those released by the U.K. reprocessing plant at Windscale

(Sellafield), particularly with respect to alpha emitters. The French plant has more advanced pollution control technology and more exacting discharge limits. The slightly lower throughput at the French plant does not account for its superior environmental performance. The comparative figures for three types of discharge are set out in table 9.8, which is taken from Taylor (1982, page 24).

Table 9.8 Cumulative Discharges in Curies for 1970 to 1976

	Cap de la Hague	Windscale (Sellafield)	Windscale/Cap de la Hague
Caesium 134	2,580	99,600	39 times
Caesium 137	15,111	489,673	32 times
Alpha	62	21,604	348 times

Rotblat (1982) points out that the discharge of caesium - 137 from Windscale (Sellafield) rose by 1000% over a few years due to the corrosion of fuel cladding. During the Windscale (Sellafield) Inquiry, Stott et al (1980, page 166) reported that a witness made the point that the USSR did not allow *any* transuranic discharges (transuranic elements are those with atoms heavier than those of uranium) in liquid discharges to the environment. This shows that the technology is available to remove the necessity for such environment discharges. BNFL (1982, page 5) have responded to this criticism. They state that

> 'in the UK the approach is to determine what is as low as is reasonably practicable bearing in mind the radiological significance of the discharges and the practical problems of dealing with the radioactivity retained on the site. Discharges at Cap la Hague and from Sellafield (Windscale) thus cannot be compared directly because circumstances are not the same, e.g. the job to be done, environmental factors and the age of the plant'.

BNFL appear to be saying that if the plant is old or the location of the plant makes it difficult to limit emissions then this justifies higher discharges. There is no mention of the consequences for human health of these higher discharges.

The U.S. EPA standards, if applied in the U.K., would restrict the discharge of alpha emitters from the new reprocessing plant at Windscale (Sellafield) to 15 millicuries per year and even less for the older Magnox plant, according to Wynne (1978). At present the Magnox reactor at Windscale (Sellafield) is discharging about 2,500 curies of alpha emitters per year and the new reprocessing plant is planned to discharge about 360 curies per year. This means that alpha emissions from the two plants will be roughly 200,000 times greater than the U.S. EPA standard, and Dr.

Karl Morgan has observed that the Magnox reactor would be closed down on safety grounds if it were located in the U.S.A. In their report, the Environment Committee (1986, page 99) concluded that 'Sellafield has become a by-word for the dirty end of the industry in the nuclear world. The international colleagues of Sellafield's scientists and engineers appeared to be embarrassed.'

In view of the ignorance of environmental pathways (see chapter 10), particularly in the 1950s, it is alarming to discover that in its early days Windscale (Sellafield) deliberately pumped out extra radioactive material into the Irish Sea. In 1958, John Dunster (who subsequently became director of the NRPB) stated that

> 'discharges from Windscale (Sellafield) have been deliberately maintained... high enough to obtain detectable levels in samples of fish, seaweed and shore sand, and the experiment is still proceeding. In 1956 the rate of discharge of radioactivity was deliberately increased partly to dispose of unwanted wastes, but principally to yield better experimental data',

quoted by Jones (1984). In response to this quotation by Jones, Dunster has replied that the discharges were below the authorized limits, had government support and aimed to establish limits on the discharges at an early enough time to allow plant modifications to be made if necessary, Jones (1984). Thus Dunster has confirmed that in the 1950s the radioactive discharges from Windscale (Sellafield) were deliberately increased as an irreversible experiment with the environment. Such action now would result in a criminal prosecution.

It has been found that grazing animals on the Cumbrian coast have levels of radioactive material very far above those in control animals. These results, which are taken from Popplewell et al (1981) and reproduced by Taylor (1982, page 39), are set our in table 9.9.

This demonstrates that radioactivity is entering the food chain in Cumbria. Taylor (1982, page 39) suggests the source of these high levels of plutonium is the marine discharges of radioactive material by the Windscale (Sellafield) plant and that somehow the radioactive material has come ashore to be ingested by grazing animals. As explained in chapter 10, there is insufficient knowledge to make reliable estimates of the way radioactivity concentrates in the food chain. Consequently it is not possible to make accurate predictions of the resulting human dose. However, as table 9.9 illustrates, the food chain can result in some remarkably large concentrations of radioactive material with the result that an initially small environmental emission may yield an ultimately lethal dose to humans.

Between the 11th and 16th November, 1983, during annual maintenance of the reprocessing plant, some radioactive liquid was discharged into the Irish Sea. On the 11th November the environmental pressure group Greenpeace discovered a radioactive slick, and by the 19th November local

Table 9.9 Radioactivity in the Liver*

		Windscale	Ravenglass	Control
Cow	Plutonium 238,239&240	290	630	1
	Americium 241	230	370	2
	Caesium 137	44,000	9,900	350
Sheep	Plutonium 238,239&240	100	11,500	9
	Americium 241	-	6,600	1
	Caesium 137	10,400	5,600	370

*Measured in 0.001 becquerels per kilogram (1 curie = 3.7×10^{10} bequerels).

beaches had been polluted and were closed to the public. The main contaminant was ruthenium-106, which is a beta emitter with a half life of one year. Some twenty five miles of the coast around Windscale (Sellafield) were contaminated, and in places the radiation was a thousand times higher than natural background levels. BNFL claims that about 600 curies were released, whilst Dr. Philip Day of Manchester University has said that 'it is unlikely to have been less than 3,000 curies, that is, at least five times the figure given officially', Tucker (1984). Radioactive items such as pieces of cork, rope, string, rubber and plastic were still being washed up on the beaches in late December 1983. They were sufficiently radioactive that, if placed next to a person's skin (e.g. whilst sunbathing), they would have delivered one year's permissible dose in less than an hour, Tucker (1983e). Indeed, it was not until the end of July 1984, eight and a half months after the accident, that the government announced that the beaches were safe for public use.

In August 1984, as a result of police enquiries following the radioactive discharge in November 1983, BNFL was charged with five offences under the Radioactive Substances Act (1960) and the Nuclear Installations Act (1965). The charges against BNFL were failure to keep exposures ALARA, failure to keep adequate records and failure to adequately control radioactive material on the Windscale (Sellafield) site. The prosecution of BNFL was the first test of the ALARA principle in the British courts.

On 23rd July, 1985, the Carlisle Crown Court found BNFL guilty on three charges. BNFL had previously pleaded guilty to a fourth charge. The judge imposed fines totalling £10,000 on BNFL, and also ordered them to pay the prosecution costs, up to a maximum of £60,000. BNFL was found guilty of (a) breaching the ALARA principle, (b) failure to take all reasonable steps to minimise the exposure of persons to radioactivity, (c) failure to keep adequate records of operations, and (d) failure to take all reasonable steps to minimise the exposure of persons to radioactivity, Atom (1985c).

In a statement to the House of Commons on 14th February 1984 concerning the Windscale (Sellafield) discharges in November 1983, Patrick Jenkin, the Secretary of State for the Environment said 'there is no evidence to suggest that this contamination, although very unsatisfactory, *could* cause significant damage to anyone's health', Atom (1984d). This reveals Jenkin's ignorance of some of the basic facts of radiobiology. Small doses of radiation can cause cancer and subsequent death. Perhaps Jenkin does not consider this 'significant damage'. In similar vein, Con Allday, the chairman of BNFL, has written of the November 1983 discharge that 'we can confidently claim that no harm has or will be caused to anyone', BNFL (1984, page 9).

The Windscale (Sellafield) pipeline discharges into the Irish Sea, and concern about the discharges has been expressed by Irish politicians. In 1984 the Irish deputy prime minister asked Britain to close the Windscale (Sellafield) pipeline into the Irish Sea, and in July 1984 Mr. Charles Haughey, the former Irish prime minister, called for Windscale (Sellafield) to be closed down, Lean et al (1984). He described the contamination from the plant as 'bordering on a crime against the people on both sides of the Irish Sea', Brown et al (1984).

The Environment Committee (1986, page 58) stated that

> 'Sellafield is the largest recorded source of radioactive discharge in the world and as a result the Irish Sea is the most radioactive sea in the world. We found for example that the Swedes could identify radioactive traces in fish off their coast being largely attributed to Sellafield, greater even than contamination from adjacent Swedish nuclear power stations'.

9.3.2 Health Effects of Windscale (Sellafield)

Geary et al (1979) compared the rate of myloid leukaemia in Lancashire (excluding Ormskirk) for 1965 to 1970 with that for 1971 to 1976. They found that in the second six year period the rate of myeloid leukaemia was 86% higher. This result applied to each of the five health districts in Lancashire. Geary et al (1979) suggested that a possible cause of this considerable increase in myeloid leukaemia might be the increased exposure of the population to radiation e.g. the deposits of plutonium and americium in the Ravenglass Estuary.

Baverstock et al (1979) pointed out that the doubling of myeloid leukaemia rates may have been part of a national increase. Leck et al

(1980) investigated this possibility and, differentiating between acute and chronic myeloid leukaemia, their results are summarized in table 9.10. This also shows that for both types of leukaemia the increase in Lancashire was well above the national increase and was markedly higher for chronic myeloid laukaemia.

Table 9.10 Increases in Myloid Leukaemia*

	Acute ML	Chronic ML	Total ML
Lancashire	+73%	+52%	+61%
England & Wales	+47%	+ 9%	+36%

*1966 to 1970 compared with 1971-1975.

Reid et al (1979) reported that a general practice of some 9,000 patients at Lytham St. Anns, Lancashire experienced a sharp increase in cases of myeloid leukaemia. For the 14 years 1958-1971 there was only one case, whilst the 7 years 1972-1978 produced 11 cases, a 22 fold increase. The authors rule out an increase in diagnostic awareness since other types of cancer such as myeloma and lymphoma showed little increase.

Birch et al (1979) studied the incidence of acute lymphoid leukaemia for 1954-1977 amongst children up to the age of 14 living in the North Western Regional Health Authority area. A statistically significant increase in the incidence of acute lymphoid leukaemia occurred from 1970 to 1977. The authors rule out changes in diagnosis as the cause of the increase in cancer rates and suggest environmental factors may have been the cause. The years for which Reid et al (1979) and Birch et al (1979) have found higher leukaemia rates are also the years for which Geary et al (1979) found an increase in leukaemia rates.

Gorst et al (1984) have studied the recent incidence of acute leukaemia in the North West Regional Health Authority area amongst those over 14 years of age. Over a two year period they found that for Lancaster district there was a clear excess of cases, see table 9.11.

Table 9.11 Number of Acute Leukaemias Over Age 14 per Million per Year - Lancaster District

Year	Observed	Expected	O/E
1982-83	10	3.4	2.94
1983-84	9	3.7	2.43

The high leukaemia rate continued into 1984-85, and in the first 17 weeks of this year there were 5 cases i.e. 15.3 per year.

Urquhart (1983) contains the following table which shows that, in the area served by the Cumbria Area Health Authority, the leukaemia rate in the 1970s was high and rising.

Table 9.12 Leukaemias in Cumbria, 1968-1978

	Observed	Expected	Observed/ Expected
RES Neoplasms			
1968-73	101	88.2	1.15
1974-78	95	80.4	1.18
Multiple Myelomas			
1968-73	12	15.6	0.77
1974-78	34	13.5	2.52

A report by PERG (1980) presents data showing that the standardized mortality rates for leukaemia for 1969-1973 in Cumberland (now called Cumbria) were 20% higher for both males and females than the average for England and Wales. However, it is later pointed out that this is not a statistically significant difference. PERG (1980) also provided crude death rates from leukaemia for the years 1963-1975, unadjusted for age and sex, for the Barrow distict which is south west of Windscale (Sellafield). For the 13 year period the death rate from leukaemia in the Barrow district was 25% higher than that for England and Wales over the same period. Finally PERG (1980) produced data on standard mortality rates for all types of cancer in south west Cumbria for the years 1975 to 1978. Compared with the whole of England and Wales, the cancer rate in south west Cumbria was 74% higher for males and 42% higher for females. It is noteworthy that a PERG investigator was refused access to data on thyroid cancer held in Barrow on the grounds that the information was "politically sensitive".

The suggestion of a link between the increase in myeloid leukaemia in Lancashire and radioactive discharges from Windscale (Sellafield) has been rejected by the chief medical officer of the Windscale (Sellafield) plant, Schofield (1979). He argues that the measured discharges from the plant are well within the current safety standards and so cannot be responsible for the increase in cancer rates. PERG (1980) conclude that, whilst an association between radioactive material emitted by Windscale (Sellafield) and higher cancer rates in Cumbria and Lancashire has not been demonstrated, such a link cannot be ruled out, and further study is necessary.

Using the ICRP estimates of the dangers of low level radiation, Taylor (1982, page 18) calculates that the radioactive discharges from Windscale

(Sellafield) up to 1978 have caused 30 fatal cancers, 30 non-fatal cancers and about 30 genetic defects. If the BEIR risk factors are used all these figures could be up to five times higher. So, using official discharge figures and risk factors, the Windscale (Sellafield) plant is estimated to have seriously affected or killed between 90 and 450 local residents. BNFL (1982, page 5) have replied to this criticism. They suggest that 30 deaths from radiation induced cancer are not worth bothering about and that the risks from other activities such as smoking, are higher.

Mummery et al (1980) report estimates of the radioactive discharges to the environment likely to result from the operation of the new THORP plant which is currently under construction at Windscale (Sellafield). Atmospheric discharges (largely krypton-85) are estimated at 140 million curies per year and liquid discharges to the sea (largely tritium) at one million curies per year. The atmospheric discharges of krypton-85 will be seven times larger than the limit set by the U.S. EPA. Subject to satisfactory development and to being justified on a cost-benefit basis, a krypton removal system may be installed to reduce the krypton discharges by a factor of ten. For discharges of plutonium-239 and other long lived alpha emitters the atmospheric discharges will be roughly five times larger than the EPA standard. Using ICRP risk estimates Mummery et al (1980) have calculated that the normal operation of the THORP plant will lead to two cancer deaths and one serious genetic abnormality each year. In evidence to the Windscale Inquiry (Parker (1978, Appendices 3 and 4)), Sir Edward Pochin estimated that the routine discharges of the THORP plant will cause 2.2 cancer deaths and one substantial genetic defect every year. These estimates assume that a krypton removal system is not built. Both Mummery et al and Pochin have used the ICRP risk factors, and if other risk factors are used the estimated number of cancer deaths can increase appreciably.

9.3.2.1 The 1957 Fire

On 10th October, 1957, there was a fire in a reactor at Windscale (Sellafield) used for the production of plutonium which caused one of the most serious nuclear accidents anywhere in the world to date. This fire involved approximately 8 tonnes of uranium and it is estimated by Taylor (1981) that roughly 370,000 curies of radioactivity were accidentally released to the atmosphere. The amount of radioactivity, measured in man-rems released in this fire was roughly 16 to 21 times larger than in the Three Mile Island accident, Pochin (1983, page 84). The plume of radioactive material was blown from Windscale (Sellafield) over Barrow and Lancashire and eventually to London. People living in the path of this plume were exposed to radiation in four ways. External radiation occurred due to people being directly irradiated by beta and gamma rays following their immersion in the radioactive cloud and due to gamma irradiation from deposited radioactive material. Internal radiation was created by the inhalation of radioactive material and the ingestion of contaminated water and food. To try to reduce the radiation dose from contaminated food, local milk was destroyed. Herbet (1982) reports that the consumption of milk produced on farms in an area 200 square miles around Windscale (Sellafield) was banned, and the milk from these farms dumped in the sea.

Whilst the individual doses of radiation have been estimated to be low, about 20 million people in the U.K. were exposed to some extra radiation. Taylor (1981) has applied the ICRP risk factors and other higher risk factors to the estimated doses and calculated the implied adverse health effects of the Windscale (Sellafield) accident on the U.K. population. The resulting estimates of Taylor are set out in table 9.13.

Table 9.13 Adverse Health Effects of the Windscale (Sellafield) Accident of 1957

	ICRP Risk Estimates	Higher Risk Estimates
Thyroid cancers	250	2,500
Thyroid cancer deaths	12	250
Lung cancer deaths	1	10
Other cancer deaths	2	20
Hereditary effects	1	1

Table 9.13 shows that the Windscale (Sellafield) accident caused the deaths from cancer of between 15 and 280 people in the U.K.

A quarter of a century after this fire, the NRPB produced a report estimating the effects upon the UK population, Crick et al (1982). It was estimated that the average radiation doses to the thyroids of local children were between 1 and 6 rem (10 and 60 mSv), with the highest recorded dose being 16 rem (160 mSv), Crick et al (1982, page 1). Without the ban on the consumption of local milk these doses could have reached 36 rem (360 mSv). This report also contains estimates of the thyroid dose for adults living in different parts of the UK. These are:-

Cumbria	1.8 rem	(18.0 mSv)
Leeds	0.14 rem	(1.4 mSv)
London	0.032 rem	(0.32 mSv)

Crick et al have used the ICRP risk factors to produce estimates of the adverse health effects of the fire and these health effects are similar to those of Taylor (1981) which are set out in table 9.13. They estimate that the fire caused 237 non-fatal thyroid cancers, 13 fatal thyroid cancers and 7 deaths from other cancers and hereditary effects. Hence, the NRPB report estimated that the Windscale (Sellafield) fire of 1957 resulted in the death of some twenty people in the UK. Five months after this report was published, in answer to a Parliamentary question on 18th April 1983, Mrs. Thatcher said "there is no evidence that anyone in the UK had died as a consequence of the release of polonium and other radioactive isotopes into the atmosphere at the Windscale fire of 1957", Atom (1983d). Since Mrs. Thatcher knew of the NRPB report, she took an unreasonable view of what might be regarded as evidence, see chapter 4.

John Urquhart of Newcastle University has argued that the NRPB study did not consider the effects of polonium released during the fire. According to Urquhart (1983) the NRPB only revealed in March 1983 that significant quantities of polonium-210 had been released, four months after the publication of their report. Polonium is an alpha emitter with a half life of 140 days that is easily absorbed by the human body. Based upon measurements of the polonium concentrations at Harwell and Den Helder, Holland, Urquhart has estimated that some 370 curies were released in the form of a gas and that this could have caused up to one thousand deaths. This estimate is very much larger than the NRPB figure of twenty deaths.

Some concern has been aroused by the suggestion that, in addition to damaging the health of people in the UK, the Windscale (Sellafield) fire also caused adverse health effects in the Irish Republic. Sheehan et al (1983) report that six women, all of whom were pupils at the same school in Dundalk in 1957, subsequently gave birth to a child with Down's syndrome. In October 1957 each of these girls had an illness similar to influenza and was exposed to an unknown (but probably small) dose of radiation from the Windscale (Sellafield) fire. Dundalk is an eastern coastal town in the Republic of Ireland, and other studies have found an increase in both the incidence of Down's syndrome and deaths from leukaemia and other cancers for the east coast of Ireland. For example, Coffey et al (1982) found an upward trend in Down's syndrome births in the area of the Eastern Health Board (Dublin, Wicklow and Kildare). In 1953-4 there were 0.54 such children per 1,000 deliveries, whilst by 1979-80 this figure had risen to 1.91 i.e. three and a half times larger.

Paul Foot has reported in the Daily Mirror that in Maryport, a coastal village north of Windscale (Sellafield), four women born in the same street all subsequently had children with Down's syndrome. Foot also discovered that another four women living in Maryport had given birth to Down's syndrome children, Black (1984, page 33).

Despite the Windscale (Sellafield) fire of 1957 and the Three Mile Island accident in 1979, statements of the following kind were still being made in 1980. "There have never been any accidents to large electricity-producing power reactors in operation in the United Kingdom, or for that matter, elsewhere in the world, that have caused to be released to the atmosphere as much radioactivity as would make any significant impact on man or his environment", Beattie et al (1980). Perhaps Beattie et al do not consider the twenty plus deaths caused by the Windscale (Sellafield) fire to be significant.

9.3.2.2 The Nuclear Laundry and the Black Report

On 1st November, 1983, Independent Television (ITV) showed a programme entitled 'Windscale - the Nuclear Laundry' made by Yorkshire Television (YTV). The makers of this programme collected their own data on cancers amongst young people in Seascale, a coastal village next to Windscale (Sellafield). The programme claimed that over the years 1956-83 leukaemia incidence for Seascale children aged 0-9 years was ten times the national rate.

Whilst some of the studies referred to earlier in this section e.g. Reid et al (1979), Gorst et al (1984) and PERG (1980) indicated an increase in cancer rates in parts of Cumbria and Lancashire, they have been largely ignored. In contrast, there have been studies of large areas of Cumbria and Lancashire which have not shown a significant increase in the cancer rate, e.g. Black (1984, pp.22-23) and these negative results have been used to support reassurances to the local population that Windscale (Sellafield) is harmless. For example, after the showing of the YTV documentary, the district medical officer for West Cumbria said that 'overall figures for the country have never suggested a need to look more closely at the area', Cutler (1984). Of course there have been other studies of large areas of Cumbria and Lancashire which have indicated a cause for concern, e.g. Urquhart (1983), Birch et al (1979), PERG (1980), Geary et al (1979) and Leck et al (1980). Just after the YTV broadcast John Dunster, the director of the NRPB, said he did not believe that, in an intensely monitored place such as Windscale (Sellafield), Britain's radiation protection experts would have missed a connection between a high local cancer rate and proximity to the nuclear plant, Fishlock (1983). However, only seven months later, the Black report confirmed that the NRPB had indeed failed to notice 'an approximately 10-fold higher rate of leukaemia incidence in the under 10 year old population of Seascale', Black (1984, page 29). Thus the Black report confirmed the findings of the YTV programme and showed up the lack of vigilance by the NRPB.

The YTV researchers found a very large increase in the local cancer rate because they:-

(a) focused on the population of Seascale, the village closest to Windscale (Sellafield),

(b) considered leukaemia, an uncommon and radiosensitive cancer,

(c) studied only children, the age group who are most sensitive to radiation exposure, and

(d) did not rely on official cancer data, and discovered additional cancers for themselves.

The evidence of the adverse health effects in Seascale was there for anyone to find. But, as in Portsmouth, U.S.A., it was the media and not the official watchdogs who found the excess cancers. As Dunster implied, the NRPB ought to have found the Seascale excess.

All parties to the Seascale controversy seem agreed that the childhood death rate from leukaemia is about ten times the national rate. This finding is despite a small downward bias that is built into the analysis. Cumbria is predominantly a rural area and age and sex adjusted cancer rates are about 8% below the national average, Black (1984, page 22). Thererfore, using national cancer rates to calculate the expected number of cancers for studies of Seascale, biases the study against finding a significant increase. The continuing dispute is over the conclusions to be drawn from the accepted fact that Seascale has a very high death rate from childhood leukaemia. What is the explanation for this high death rate?

To many, the answer to this question is obvious - the large nuclear plant right next to the village which is known to discharge considerable amounts of radioactivity on a regular basis and to have been the site of a serious nuclear accident in 1957. Black looked for other possible carciogens in the area, but was unable to find any. Amazingly, Black failed to draw the obvious conclusion; that the increased leukaemia rate in Seascale was due to radioactivity from Windscale (Sellafield).

In terms of hypothesis testing, the theory advanced by the TV documentary is that discharges from Windscale (Sellafield) were responsible. The only alternative explanation the Black report offers is that the very high childhood leukaemia rate found in Seascale is due to chance. However, there is strong statistical evidence in favour of the YTV hypothesis. The incidence rate for lymphoid malignancy amongst children aged 0 to 14 years in Seascale is about 16 times larger than that for the region, Black (1984, page 32). This excess is statistically significant at the 0.0134% level, i.e. there is only one chance in 8,000 that the excess is a random fluctuation and 7,999 chances in 8,000 that there is some factor which has caused the childhood leukaemia rate in Seascale to increase. Seascale has the most significant excess of childhood lymphoid malignancies of any ward in the U.K., Jones (1984).

In the scientific literature, before a hypothesis is accepted, it is common to require that the chances of the hypothesis being false are less than one in twenty, i.e. the result is statistically significant at the 5% level. The hypothesis that there is an increased childhood leukaemia rate in Seascale is about 400 times more significant than this commonly accepted level. Pomiankowski (1984) has argued that, when two further cases of lymphoma diagnosed in 1983 in children living in Seascale are included, the probability that the increase in lyphoid malignancy in Seascale is due to chance drops to roughly one chance in a million! Hence it is very hard to maintain that the increased rates of certain types of childhood cancer in Seascale are due to chance.

A dispute has developed over which cases should be included in studies of cancer in West Cumbria. Cutler (1985) maintains that seven cases of cancer were excluded from the statistical analysis of cancer rates in the Black Report, even though all seven cases are listed at the start of the Black Report (pp.14-16). Craft et al (1985) and Gardner (1985) have accepted that six of these seven cases of cancer were omitted from the statistical analysis, and give reasons for their exclusion. A case of erythroleukaemia was omitted because in 1964, when the child died, the disease was not classified as leukaemia, although now such a death would be classified as leukaemia. Two of the cases were diagnosed after 1982 and were excluded because they fell outside the study period. Another two cases were excluded because, according to the regional cancer registry, they were diagnosed after their fifteenth birthday. In fact, as stated on page 16 of the Black Report, they died at the ages of 6 and 13 years and so *must* have been diagnosed before their fifteenth birthday. Finally, another case was accidentally excluded because of a mistake concerning the address. Since the total number of cancers being considered is a small number, six extra cancers represents a substantial increase. Hence, apparently arbitrary decisions about the length of the study period or what constitutes a leukaemia can have a major effect upon the results.

The Black report justifies its conclusion, that the increased leukaemia rate in Seascale is due to chance, by a very strange argument. Previously, studies showing that low level radiation is dangerous have been dismissed on the grounds that the results are not statistically significant at the 5% level, i.e. the increase in the cancer rate is too small to be reasonably sure it is not just a random fluctuation. Now, when there is a study involving low level radiation which has very highly significant results, it is dismissed on the grounds that its results are *too* significant. This logic means that if even more cancers occur in Seascale this would strengthen Black's view that chance and not Windscale (Sellafield) is the cause!

The NRPB has conducted an analysis of the problem for the Black inquiry. They estimated the additional dose of radiation from Windscale (Sellafield) to which the children living in Seascale may have been exposed and then multiplied this dose by an estimated risk factor. They concluded that radiation from Windscale (Sellafield) could only account for about 2.5% of the increase in childhood leukaemias in Seascale. Thus, the increase at Seascale was too large for it to have been caused by radiation, leaving a random fluctuation as the explanation!

This is nonsense for three reasons. First, the NRPB calculations involve estimates of two main numbers (a) the radiation dose to children from Windscale (Sellafield) and (b) the risk factor. It is the purpose of this book to argue that estimated risk factors are very imprecise, and they could easily be wrongby a factor of seven. Similarly, it is very difficult to estimate the dose to the most exposed children with any great accuracy, and again the estimates could be wrong by a factor of six. For example, Black (1984, page 74) states that the 'NRPB in their report consider that their estimates of concentrations of radionuclides in marine foods are probably accurate to within a factor of about 5 in most cases'. This imprecision means that discharges from Windscale (Sellafield) could account for 6 x 7 = 42 times as many childhood leukaemia deaths as estimated by the NRPB, i.e. 105% of the observed number.

It subsequently emerged in February 1986 that the atmospheric discharges from Windscale (Sellafield) between 1952 and 1955 were 40 times higher than stated in the Black report, Spackman et al (1986). This was revealed by physicist Dr. Derek Jakeman, who was employed at Windscale (Sellafield) in the 1950s, and investigated a large discharge in 1955. His requests for further information were refused, and he was threatened with dismissal by the UKAEA, who were then the plant operators. His unofficial report on the accident was ignored, and he resigned. BNFL have confirmed that Jakeman is correct, and that the figures supplied to the Black inquiry were a considerable understatement.

Second, the Black inquiry have rejected the possibility that part of the increase in childhood leukaemia in Seascale is caused by radiation from Windscale (Sellafield) and part is due to chance. Black preferred to attribute *all* of the increase to chance because this was thought to be more plausible.

Third, discarding data because it does not fit your theoretical model means that the theory concerned is irrefutable. Thus, the NRPB have a model for predicting Seascale cancers based upon radiation risk factors etc., and they believe it to be reasonably accurate. If there is a conflict between the predictions of the NRPB model and reality, reality must be wrong!

Despite clear evidence of a substantial increase in deaths from childhood leukaemia in Seascale, the Black inquiry concluded that 'we can give a qualified reassurance to the people who are concerned about a possible health hazard in the neighbourhood of Sellafield', Black (1984, page 93). The Black report argued that there were *too many* cases of childhood leukaemia in Seascale for radiation from Windscale (Sellafield) to be the cause! They preferred to rely on the million to one possibility that the extra deaths were due to chance. It is hard to justify these conclusions in terms of scientific methodology, and there may be other explanations as to why such conclusions were reached. According to Jones (1984) when Peter

Wilkinson, director of Greenpeace, gave evidence to the Black Inquiry, Sir Douglas Black told him that the primary function of his inquiry was 'to reassure the Cumbrian public'.

Subsequently, the conclusions of the Black inquiry have been used to justify statements to the effect that Windscale (Sellafield) has not affected the health of local inhabitants. For example, Con Allday, the chairman of BNFL has written that the Black report 'has shown this claim (that Windscale (Sellafield) has increased the local cancer rate) to be unsubstantiated', BNFL (1984). He subsequently claimed that the radioactive discharges by Windscale (Sellafield) into the Irish Sea were quite harmless, Wilkinson (1986). On 5th March 1986 Mr. Whitney, a governement spokesman, stated in the House of Commons that 'on the basis of current evidence I am advised that there is no established causal link between nuclear power installations and the incidence of disease in populations near to them', Atom (1986). Such statements are only possible because of the very strange conclusions which the Black inquiry drew from the evidence.

One leading politician has expressed forthright views about the Black report. Mr. Haughhey, the former Irish prime minister, described it as a dreadful piece of whitewash and said that

> 'if there is a high incidence of leukaemia in an area where a nuclear plant is situated, surely to God the obvious interpretation is that the plant was responsible for it. These figures alone would, in my view, justify closing down the plant immediately for further investigation and certainly putting a lot of people in gaol who have clearly been telling us lies over the past four or five years about this matter. The situation is such that the people responsible for that particular plant and the denials should be put in gaol',

Brown et al (1984). In July 1986 the government of the Isle of Man also called for the closure of Windscale (Sellafield), Hamilton (1986).

9.4 Three Mile Island

On 28th March, 1979, an accident at the pressurised water nuclear reactor at Three Mile Island at Middletown, Pennsylvania, resulted in the release of an unknown quantity of radioactive material. The dose per person probably did not exceed a few rem, but the monitor devices 'peaked' during the accident so accurate measurements are unavailable. MacLeod (1981) reports that in the nine months before the accident the Pennsylvanian State Health Department found 17 cases of depressed thyroid function amongst newly born children. In the nine months after the accident the number of cases of depressed thyroid function jumped to 27, with the increase being concentrated in the eastern part of Pennsylvania downwind of the Three Mile Island reactor. MacLeod (1981) concluded that an epidemiological study was warranted to try to determine whether radioactive iodine from the Three Mile Island reactor caused a significant increase in hypothroidism.

Sternglass (1981, page 228) estimated that the accident led to the relase of 10 million curies of radiation into the environment. Sternglass (1981, pp.233-237) has collected data on infant mortality in the areas around the Three Mile Island plant for March 1979 and May 1979 and calculated the percentage changes in infant mortality for different areas which are set out in table 9.14.

9.14 Changes in Infant Mortality Between March and May 1979

New York State (excl. New York City)	+ 52%
Pennsylvania State	+ 44%
Maryland	+ 26%
New Jersey State	+ 9%
Ohio State	0%
United States	- 11%
New York City	- 22%

The largest releases of radiation occurred when the wind was blowing north and west, some smaller releases took place when the wind was in a southerly direction, whilst little radiation was released when the wind was blowing eastwards. Sternglass concludes from the pattern of changes in infant mortality that radiation emissions from the Three Mile Island plant at the end of March 1979 were responsible for an increase in local infant mortality. Sternglass (1981, pp.236-237) also attempted to obtain data on infant mortality in the immediate vicinity of the plant where the increase was expected to be the highest, but he was unable to get data on a county basis for Pennsylvania. However, an executive at the Harrisburg Hospital supplied him with the average infant mortality rate for the hospital for February, March, April and May 1979. The figures showed that the rate for May 1979 was 323% higher than for the preceding 3 months (which were before the accident). Following the Three Mile Island accident, there was an increase in Down's syndrome births in Tompkins County, which is some 200 miles to the north of the reactor. There was also an increase in Down's syndrome births in Broome and Chemung counties for 1980 and 1981, Bertell (1985, page 205).

Sternglass (1981, page 247) speculates that the accident may lead to between 4,000 and 8,000 extra cancer deaths in addition to the rise in infant mortality. Gofman et al (1979, page x) state that the Three Mile Island accident caused at least six deaths, but that the number of extra deaths could easily be 60 to 70 cancer deaths. These conclusions are in sharp contrast to the official report, the Kemeny report, which concluded that the only health effects of the accident were psychological! It is this latter view that is reiterated in publications. Thus a pamphlet published by the UKAEA states that 'no employee or member of the public was injured as a result of the accident', Nuclear Power Information Group (1983).

9.5 Aldermaston and Burghfield

The AWRE, Aldermaston, discharges radioactive material down a pipeline into the River Thames at Pangbourne and into the atmosphere. The Royal Ordnance Factory at Burghfield, which assembles the U.K.'s nuclear warheads, discharges such material into the River Kennet (and thereby the Thames) and the atmosphere.

On 3rd December, 1985, a documentary entitled 'Inside Britain's Bomb', made by the same YTV team who made 'Windscale Nuclear Laundry', was shown by ITV. This programme presented what is so far the only study of the effects of these two MoD establishments, where Britain's bomb is designed, manufactured and assembled, on the health of the local population.

The area examined by YTV was the nine rural wards which form two circles of roughly 2.5 miles radius around each nuclear plant. The YTV team found that for the period 1971 to 1985, for people aged under 25 there were 11 cancers amongst those living around Aldermaston, and 8 cancers amongst those living around Burghfield. Of these 19 cancers, 15 were leukaemia and lymphatic cancers, which are 'radiosensitive' cancers. It was also found that the leukaemia death rate for children aged under 10 was five times the national average, and the cancer death rate for children aged under 10 was three times the national average. For just Burghfield parish the incidence of leukaemia and lymphatic cancer amongst children under 5 years old was ten times the national average.

Subsequently Urquhart et al (1986) published the data on the incidence of leukaemia and lymphatic cancers relative to the national rates.

Table 9.15 Relative Increase in the Incidence of Leukaemia and Lymphatic Cancers

Area	0-4	0-14	0-24
Aldermaston & Burghfield circles	+280%	+90%	+80%
Burghfield circle	+560%	+210%	+130%
Burghfield Parish	+890%	+310%	+200%

A study of the incidence of childhood leukaemia in West Berkshire, which includes Aldermaston and Burghfield, has been conducted by Barton et al (1985). They studied children aged under 10 years for the years 1972 to 1984 and found that leukaemia incidence was 65% above the national rate for children under 5 years old. This result was statistically significant at the 1% level. The lower increase in childhood cancer rates in West Berkshire is consistent with the higher rates found by YTV for the Aldermaston/Burghfield area. Large parts of West Berkshire are upstream and upwind of the two nuclear plants, and so the health effects are expected to be less.

With the publicity given to the radiation risks from Aldermaston/Burghfield, John Bradshaw, a former Concorde pilot, has revealed some new information. When flying into Heathrow airport from the west, the radiation meter in Concorde would often swing into the amber warning zone. It became apparent that this warning occurred as the plane flew over the Aldermaston/Burghfield area at a height of 10,000 feet (about two miles), Handley (1986).

9.6 Chernobyl

On Saturday 26th April 1986 at 1.23 a.m. there was a major accident in one of the four nuclear reactors at Chernobyl in the Ukraine. There was an explosion followed by a fire, and this resulted in the release of a considerable amount of radioactive material to the atmosphere. Two workers were killed by the initial explosion, and many of the firemen who fought the blaze received lethal doses of radiation. Six days after the accident, the radiation level in the vicinity of the reactor was about 190 rem (1,900 mSv) per hour. Dr. Andrei Voronyov said that an examination of people living in the settlement of Pripyat close to the plant indicated that they had received a maximum dose of 50 rad (500 mGy). By July 1986 28 people had died as a direct result of this accident, another 300 had been hospitalised and 100,000 evacuated from their homes. The radioactive cloud from Chernobyl was blown across much of eastern and western Europe.

The Russians failed to alert neighbouring countries to the imminent danger from this cloud. Two days after the accident, when increased levels of radiation were measured in Sweden, the Russians initially denied that any nuclear accident had occurred. They delayed evacuating the zone 18 miles around the plant until a week after the accident, and some of the Soviet press coverage of the accident was intentionally misleading. It appears that the Russians would have liked to conceal this accident, as they did the explosion in the Urals in 1957-58.

Something like a million times more radioactivity was released at Chernobyl than in the Three Mile accident of 1979. While the radiation levels immediately around the plant were deadly, the doses to people living in other countries carried much less risk. Although small in absolute terms, some very large proportionate increases were produced by the radiation released from Chernobyl. For example, in Poland the quantity of radioactive iodine in the air was 500 times the normal amount, and in northern England the milk was 200 times more radioactive than usual. While these increases are alarming, the extra radiation doses to UK residents are estimated to have been tiny. The UKAEA has calculated that a one year old child living in the north of England will have received a dose of approxiamtely 0.1 rem (1 mSv), Gittus (1986).

Of course all radiation exposure carries the risk of death, and the exposure of many millions of people implies a non-trivial number of deaths. Professor Herbert Begemann, a West German haematologist, has estimated that in East Germany alone the radiation dose from Chernobyl will produce 600 to 2,000 extra cancers *per year*. This elevated cancer rate could last for 20 to 30 years. Presumably, the worst adverse health effects will appear over the next 40 years amongst the Russian population. In view of this, arrangements have been made by the Russian authorities to follow up the health of the 100,000 evacuees for the rest of their lives.

On 6th May 1986 Kenneth Baker, the Secretary of State for the Environment, said that radioactivity in Britain is 'nowhere near the levels at which there is any hazard to health'. On the same day John Dunster, the director of the NRPB, said that a few tens of people in Britain might die of cancer over the next fifty years. In this conflict between the minister and one of his senior advisors, it is Dunster who is being more realistic. Baker's statement is based on the discredited notion of a threshold dose. It is curious that in his next sentence after denying there was any health hazard, Baker announced a ban on drinking rainwater in northern England and Wales and Scotland because of the dangerous levels of radioactivity.

Because of their radioactive contamination, the European Community placed a temporary ban on the importation of foodstuffs from eastern Europe on 10th May 1986. The British government set an 'action level' for meat contamination of 1,000 becquerels per kilogram. On 14th May 1986 eleven out of twelve samples from Cumbrian sheep exceeded 600 becquerels per kilogram, the level at which the European Community was banning imports from eastern Europe, and five exceeded the UK government's action level. One sample contained 2,450 becquerels per kilogram. Despite their action level being breached, the UK government failed to act. It was only five weeks later, on 20th June 1986, that the UK government imposed a ban on the movement and slaughter of sheep in north Wales and Cumbria. In the interim meat had been on sale to the public with levels of radioactive contamination which exceeded the government's action level. In one case a radiation level of 4,000 becquerels per kilogram was found in a lamb that was already on sale to the public. In Norway reindeer meat was found to contain up to 24,000 becquerels per kilogram.

In 1984 Peter Neporozhny, the Soviet Minister for Power and Electrification, said that nuclear power stations 'are very economical and can be built in the immediate vicinity of a city because they do not emit smoke and are totally safe'. The Chernobyl disaster shows that not only are nuclear reactors not totally safe, but nuclear accidents do not just affect their immediate vicinity. The radioactive plume from Chernobyl contaminated most of Europe, and so the safety of a particular country's reactors is a matter of international importance.

9.7 Two U.S. Nuclear Reactors and Cancer

In Shutdown (1979, pp.124-125) Sternglass reported the effects of the Shippingport pressurised water reactor, opened in 1958 in Beaver County, Pennsylvania, on deaths from cancer in the surrounding area. After first discovering elevated levels of radiation in the local environment, he found that for Beaver County the cancer mortality rate for 1958 to 1968 showed an increase of 39%, whilst the corresponding increase for the U.S.A. was only 8%. As distance from the Shippingport reactor increased so the increase in cancer rates was lower. The analysis by Sternglass of the increase in cancer deaths (and infant mortality) in the area surrounding the Shippingport reactor is summarized by Lewis (1973). Sternglass in Shutdown (1979, page 127) also reported similar results for the Millstone reactor in Waterford, Connecticut. Between 1970, the year the reactor started, and 1975, cancer mortality in Waterford increased by 58% as against only 6% for the U.S.A. Again the cancer effect decreased with distance from the reactor. Whilst these results are not conclusive evidence that low levels of radioactive emissions are responsible for the increases in cancer rates, it is a highly plausible explanation.

9.8 Five U.S. Nuclear Reactors and Infant Mortality

DeGroot (1972) studied the effects of low level radiation emissions from four U.S. nuclear reactors on local infant mortality rates. The reactors studied were Dresden (Illinois), Shippingport (Pennsylvania), Indian Point (New York) and Brookhaven (New York). DeGroot used time series multiple regression analysis for 18 annual observations. Apart from Shippingport, DeGroot's simplistic analysis suggested a weak positive correlation between infant mortality in the immediate vicinity of the plant and the level of radioactive emissions. In Shutdown (1979, page 129) and Sternglass (1981, pp. 145 and 151) Sternglass pointed out that the lack of any suggestion of correlation at Shippingport may have been due to inaccurate data on the emissions of radioactive material. DeGroot accepted that his approach was simple and his results inconclusive, but thought there was sufficient indication of a connection between radioactive emissions and infant mortality for more detailed studies to be carried out.

Sternglass (1972 and 1974) gave details of the results of investigations he conducted into the health effects of three U.S. nuclear reactors; Hanford, Dresden and Indian Point. Sternglass looked at the effects of these reactors on local infant mortality rates. Unlike cancer, where there is a long latent period, the effects of radioactivity on infant mortality appear to be fairly rapid. In consequence, Sternglass did not use any time lag between radioactive emissions and infant mortality. The Hanford plant in Benton County, Washington, was opened in 1944 and the rate of infant mortality for Benton County was 150% higher in 1945 than in 1943, suggesting that the opening of the plant had caused this increase. In support of this interpretation is the fact that for more distant counties the effect upon infant mortality rates decreased until, for counties over 100 miles away, there was no effect.

The Dresden reactor was opened in Grundy County, Illinois in 1960. For the period 1959 ro 1968 Sternglass regressed the excess infant mortality in Illinois relative to the control state of Ohio, on the annual average gaseous radioactive discharges from the Dresden reactor. The correlation coefficient was 0.86 and the t value 4.6. This shows a highly significant positive correlation between radioactive discharges and infant mortality in Illinois. Sternglass then compared the changes in infant mortality in the six counties immediately next to the reactor with that in six counties in Illinois more than 40 miles to the west of the reactor. Over the period 1964 to 1966 the infant mortality rate increased by 48% in the surrounding counties and decreased by 2% in the control countries. Sternglass appears to have chosen the years 1964 to 1966 because over this period emissions rose to their peak level. He also reports that over the period 1964 to 1966 there was an increase in premature births in Grundy county of 140%, but little change in the six control counties. Even in the year of peak emission (1966) the measured external doses were only 0.07 to 0.08 rad (0.7 to 0.8 mGy), and so Sternglass has found evidence suggesting that *very* low levels of radioactive emissions can have an adverse effect on infant mortality rates and also cause premature births.

The pressurised water reactor at Indian Point, Westchester county, New York, was opened in 1962. For the years 1963 to 1969 Sternglass regressed excess infant mortality for Westchester and Rockland counties relative to a control county, Nassau, on the two year average annual amounts of radoactive liquid waste discharged from Indian Point (expressed as a percentage of the permissible limit). The correlation coefficient was 0.97 whilst the estimated slope was 0.38 with a t value of 9.7. This estimated relationship between dose and response is positive at a very high level of statistical significance. Again this demonstrates that radioactive doses well within the permissible levels can lead to a considerable increase in local infant mortality rates. Sternglass repeated the analysis but substituted four upstate control counties for Nassau county. This time the correlation coefficient was 0.96, the slope was 0.66 and the t value was 7.4. Again the positive slope coefficient was highly statistically significant. Hence the previous result is confirmed using a different control. Sternglass estimated that for the year of peak emissions (1966) infant mortality in Westchester and Rockland counties was increased by 41% with a lesser effect for more distant counties, and that the absolute number of extra infant deaths in 1966 as a result of the radioactive emissions from the Indian Point reactor was over 850.

9.9 Various Nuclear Establishments

Ryle et al (1980) report that a study by the Bremen Institute for Biological Safety of the population around the Lingen nuclear reactor in West Germany, found that 10 years prior to the operation of the plant there were 30 leukaemia deaths, whilst in the 10 years after nuclear operations began (1968-1978) the same population suffered 230 deaths from leukaemia, of which 170 were children under 15 years old. According to Ryle et al (1980) the Getty Nuclear Fuel services plant at Jonesboro, Tennessee has been shown to have caused a 95% increase in the death rate from cancer throughout the surrounding county during the 20 years of its operation.

The Big Rock Nuclear reactor, three miles north west of Charlevoix, Michigan, began operation in 1962. A local doctor, Dr. Gerald Drake, became concerned about the effects upon local health, and in 1974 he completed a study of the effects of the first ten years of operation of this plant. He found that the number of low birth weight infants (under 5.5 lbs) in Charlevoix County rose by 21% whilst the corresponding figure for the state of Michigan rose by only 1%. This result is significant at the 0.5% level of confidence. Dr. Drake also found that the increase in the rate of cancer deaths in Charlevoix County was seven times larger than that for the state. This result was satistically significant at the 0.1% level, Bertell (1985, pp.206-207).

Dounreay, on the north coast of Scotland, has been the site of nuclear research and development since 1955. The UKAEA operates a prototype fast nuclear reactor and a fast reactor fuel reprocessing plant. There is a pipeline for discharging low level waste into the Pentland Firth. Heasman et al (1986) studied cancer incidence in the area around Dounreay for the period 1968-1984 amongst people aged under 25. They found that the leukaemia rate for the area within 7.75 miles of Dounreay for 1979-84 was 9.7 times higher than the rate for Scotland as a whole.

A study of the cancer rates of members of the public living in Caithness and Sutherland between 1958 and 1982 has been conducted by the UKAEA, Wood and Smith (1986). An analysis of the crude death rates by parish, i.e. unadjusted for age differences, was carried out for deaths from lymphatic and haematopoietic cancers. A significant excess of such cancer deaths was found for the parish of Latheron in Caithness, south of Dounreay. The leukaemia rate was three times larger than expected, and the odds against finding such a result by chance are 200 to 1.

The small Suffolk town of Leiston is just over a mile away from the existing Sizewell A nuclear power station. Within a period of fifteen years a cluster of eight cases of leukaemia has occurred in a small residential area of Leiston. This cluster may be chance or there may be some causative factor e.g. Sizewell A, Financial Times (1983c).

9.10 Summary

Despite the number of nuclear establishments in the world that have been operating for many years, it is surprising there have been so few studies of the effects upon the health of the local population. For example, whilst it is known that there are considerable radioactive emissions from Windscale (Sellafield), it was only in 1984 that there was a study of the effects of Windscale (Sellafield) upon the health of the local population. It is noticeable that only two of the health studies reported in this chapter have been conducted by official organizations; the Black report and the NRPB study of the Windscale (Sellafield) fire. In each case the official study was prompted by a previous non-governmental study which showed an increase in the cancer death rate. It appears that governments do not want any such studies carried out in case they find the actual levels of emissions from nuclear establishments are causing significant increases in cancer and infant mortality. Indeed, this is what the studies in this chapter indicate.

Despite the lack of publicity given to the Windscale (Sellafield) fire of 1957, it was one of the most serious nuclear accidents to date. For example, it was an order of magnitude more serious than the Three Mile Island accident of 1979. From time to time Windscale (Sellafield) has also had other accidental releases, as in November, 1983. In addition to these accidents, Windscale (Sellafield) is one of the dirtiest nuclear plants in the world regularly discharging relatively very large amounts of radioactive material into the environment. The evidence also indicates that the Rocky Flats plant in the U.S. is causing many deaths among the local population. Thus the populations living near nuclear installations can be involuntarily, and possibly unknowingly, exposed to potentially lethal doses of radiation.

The evidence available shows that nuclear establishments are deadly. It should be stressed that dramatic accidents are not necessary for deaths to be caused amongst the local population. The studies reported here (excluding Chernobyl, Three Mile Island and the 1957 Windscale (Sellafield) accident) relate to the normal functioning of these nuclear establishments.

10 Public exposure to radiation from other sources

10.1 Introduction

The nuclear industry is sometimes referred to as a 'clean' industry. All that this means is that it does not produce unpleasant smells or grimy smoke. But this industry does produce an odourless, tasteless, invisible waste which cannot be detected by the human senses, but which is nevertheless deadly.

The nuclear industry generates radioactive material that is released to pollute the environment in one of three forms - solid, liquid and atmospheric waste. The routine operation of the nuclear industry involves the deliberate discharge of radioactive material into the air and water, and the dumping of solid nuclear waste in the ground and in the sea. In addition to this there are accidental discharges of radioactive material into the environment. When considering the health risks to the surrounding population from nuclear establishments there are two key and contentious issues, (a) how much radiation is being emitted, intentionally and unintentionally, by the nuclear establishment and (b) what are the health risks resulting from a specified level of radiation? Both these pieces of information are necessary to establish the health risks of a nuclear establishment.

In the previous chapter the health effects of the atmospheric and liquid discharges of radioactive material by nuclear establishments were considered. In this chapter, after discussing the environmental pathways by which radioactive discharges into the environment affect people, the dangers resulting from the disposal of solid waste will be considered. The pumping of liquid waste into the sea by nuclear establishments was considered in chapter 9, and in this chapter attention will be given to the dumping of solid waste in the sea. There then follows a discussion of the revised estimates of the risks of drinking water contaminated with radioactive material. Whilst there are various forms of solid waste, the only one on which there appears to be any evidence of the health effects is uranium tailings. So, after a discussion of gut uptake factors, the alarming health effects that may be produced by uranium tailings will be presented. The adverse effects of a nuclear accident in the USSR thought to have been caused by radioactive waste will be analysed. Before discussing the effects of natural radiation on health, the evidence that radiation in fertilizers is responsible for cigarettes causing lung cancer will be considered. Finally, some accidental exposures of the public are described.

10.2 Environmental Pathways

The health risks from radioactive discharges depend not only on the quantity of these discharges but also on the way in which they are concentrated in environmental pathways back to people. The aim is to check all the environmental pathways back to people in order to find the critical group, i.e. that group which receives the largest dose from all pathways combined. The discharge limits are then set with reference to the dose of radiation received by the critical group. The RCEP (1976, page 28) point out that the environmental pathway of greatest radiological significance may change over time, and indeed this appears to have happened with respect to Windscale (Sellafield). The critical group changed from those who ate laverbread made from Cumbrian seaweed to salmon fishermen in the Ravenglass estuary. Since plutonium released into the environment will remain dangerous for many thousands of years, it is impossible to predict all of the environmental pathways that may develop in the future. Hence, even if the risks from the existing environmental pathways are known, new and potentially more dangerous pathways may emerge. This implies that considerable caution should be exercised in setting discharge limits.

It appears that knowledge of the environmental pathways of radiation back to people is poor. In Stott et al (1980, pp. 160-161) it is stated that a recent Ministry of Agriculture Fisheries and Food report found that the levels of radiation in human beings as a consequence of eating fish were double the maximum predicted level. It is also stated by Stott et al (1980, page 161) that in his evidence to the Windscale (Sellafield) enquiry Taylor concluded that, even for the best researched pathway, the level of accuracy was poor. Blair et al (1974) also conclude that "the information available on the environmental distribution and redistribution of plutonium is inadequate for predicting the probable accumulation of plutonium in man". Laboratory studies of plutonium have dealt with simple chemical forms of plutonium, whilst in the environment it can form many complex organic compounds whose behaviour is difficult to predict. Thus the amount of radiation transferred from one animal or plant to another along food chains to man may be quite different for different chemical forms of plutonium, Stott et al (1980, page 169). An example is given in this chapter where the effects of chlorinated water, as opposed to pure water, on the absorption of plutonium by humans, is considered. Despite the obvious importance of a good knowledge of environmental pathways and the resulting effects upon populations, Stott et al (1980, page 151) report that the local population in the Windscale (Sellafield) area have not been monitored for their actual uptake of radiation. Thus the considerable uncertainty over environmental pathways adds an extra factor to calculating the risks from environmental emissions of low level radiation. The release of radioactive material to the environment will probably involve a combination of many different types of radioactive material e.g. strontium, iodine, caesium, plutonium, americium and tritium. Since the various radionuclides have different physical and chemical properties, each will have a different environmental pathway back to humans. So, in order to estimate the health effects of such a release, it is necessary to study many different environmental pathways. Therefore, as Gofman (1983, pp.323-328) pointed out, it is extremely difficult to calculate the total additional dose to humans.

In Shutdown (1979, pp.135-136) Gofman summarises a report by fourteen scientists at the University of Heidelberg on the doses of radiation received by people living near a pressurized water reactor in Germany. The German scientists concluded that the calculated doses were underestimated by between 100 and 1000 times, and that this underestimation was due to a systematic bias in the selection of parameters by the ICRP and other organizations since, when estimating the extent to which radiation would be transmitted back to people along environmental pathways, the ICRP had always used the lowest available estimate. For example, if there were about five or ten different experiments published estimating how much strontium 90 gets from the soil to plants they would always pick the lowest value, and they would pick the lowest value for how much strontium 90 gets from plants to beef cattle and from beef cattle to humans. The implication of this is that the doses of radiation actually received by the populations living around nuclear establishments may be many times higher than the official estimates since the official estimates represent the absolute minima.

10.3 Sea Dumping of Solid Radioactive Waste

The UK has been dumping low level solid radioactive waste at sea since 1949. The site for most of the UK's early dumps was the Hurd Deep in the English Channel and the use of this site ceased in 1963. About 15,300 tonnes of radioactive waste was dumped at the Hurd Deep, Holliday (1984, page 11). From 1955 the U.K. has increasingly switched to dumping in the Atlantic, British Nuclear Forum (1983). Since 1954 the U.K. has dumped 70,847 tonnes of solid radioactive waste in the Atlantic and this contained 16,883 curies of alpha emitters and 930,930 curies of beta and gamma emitters, Atom (1986b). In total the U.K. has dumped 73,530 tonnes in the Atlantic, Holliday (1984, page 16). As can be seen from table 10.1, the level of sea dumping by the U.K. steadily increased over time until 1982.

The 'benefit of the doubt' has been used by the U.K. government to justify continuing the sea dumping programme. In answer to a Parliamentary question in October 1983, Mr. Waldegrave said that 'in the absence of authoritative scientific advice that sea disposal within the terms of the London Convention poses a real risk to human health, or is likely to cause significant and permanent damage to the environment, this disposal method continues to form part of the national strategy for radioactive waste management', Atom (1984b). Thus the burden of proof is that sea dumping be shown to be dangerous rather than that it be shown to be safe.

In 1982 Taylor (1982, page 56) wrote that the 'dumping programme is to proceed with only the most sketchy of models of impact, and detailed impacts will be predicted upon the basis of data derived from the dumping programme'. The UK's policy is later described by Taylor (1982, page 58) as 'dump first, study later'. One rationale for sea dumping appears to be that the ocean floor is largely isolated from the human environment. However, apart from the fact that the water may circulate, evidence is now appearing of food chains linking the animals and plants of the ocean depths with those in the upper levels, including fish caught for human consumption. Taylor points out that the Atlantic dump site has not been subject to a detailed environmental impact analysis. Hence, whilst there is no evidence that this sea dumping is dangerous, there is also no evidence that it is safe.

There are theoretical models of the environmental pathways by which sea dumping leads to an increased radiation dose to humans. These predict that the maximum dose does not occur until about one hundred years after the dumping, Holliday (1984, page 45). Thus, the public exposure shortly after dumping may be insignificant, but a century later the dose may be very much larger. This means that observations for a very long period are required before the theoretical predictions of radiation exposure can be fully tested, and no one can be sure now what will be the maximum dose to the

Table 10.1 Sea Dumping in the Atlantic by the UK

Year	Tonnes	Curies Alpha	Curies Beta & Gamma
1954	0	0	0
1955	2617	47	77
1956	1038	44	33
1957	5941	1064	969
1958	3705	753	1142
1959	1198	4	74
1960	2551	74	218
1961	6327	583	1938
1962	1697	22	239
1963	7352	371	7115
1964	4392	444	15090
1965	1759	114	13754
1966	1044	78	2742
1967	722	91	1682
1968	3164	731	74837
1969	1878	390	17590
1970	1674	233	20224
1971	1434	323	8615
1972	1885	674	19049
1973	1453	739	11641
1974	1256	399	94126
1975	1350	704	52481
1976	2269	789	49777
1977	2140	930	74830
1978	2080	814	69307
1979	2014	1381	81080
1980	2693	1791	106079
1981	2517	2032	104709
1982	2697	1264	101512
1983	0	0	0
1984	0	0	0
1985	0	0	0
Totals	70847	16883	930930

public from the material that has already been dumped. It is the view of the Holliday inquiry that 'the total health detriment is uncertain', Holliday (1984, page 51). Despite this, the Nuclear Industry Radioactive Waste

Executive (NIREX) (1983) confidently stated that sea dumping 'causes no hazard to the people of any country or to the marine environment'.

An alternative view of sea dumping is that it relies on a dilute and disperse philosophy, rather than on a concentrate and isolate approach. In evidence to the Environment Committee(1986, page 38) Dr. Lewis Roberts, chairman of the UKAEA, said that 'the essential safety factor in protecting man is the enormous dilution effect of the sea itself and that is a quite different philosophy from disposal on land where you need to retain the activity in that precise location for as long as you can'. In other words, the aim of sea dumping is to spread the radiation throughout the environment.

Sea dumping is controlled by two international agreements, the London Dumping Convention (LDC) (signed by the UK in 1975) and the NEA of the OECD. The dumping site for European nuclear waste is roughly 500 miles south west of Ireland and 430 miles north west of Spain (see figure 10.1). Since 1977 only four countries have engaged in sea dumping: U.K., Belgium, Switzerland and the Netherlands, and in 1983 the Dutch announced a moritorium on sea dumping. The UK has been responsible for most of the world's sea dumping, and according to Taylor (1982, pp.60-62) over 90% of the waste dumped in the Atlantic was dumped by the UK.

On 18th February 1983, by 19 votes to 6, the signatories of the LDC passed a resolution calling for a two year suspension of sea dumping pending the outcome of an international investigation into the environmental effects. The UK voted against the motion, and since the motion was only advisory, planned to continue its sea dumping programme. According to Mr. Buchanan-Smith, M.P., the U.K. government opposed the motion because 'it called upon member states to act in advance of receiving authoritative scientific advice', Atom (1983c).

In contravention of the moratorium, the U.K. government authorised the dumping of 3,500 tonnes of radioactive waste in the summer of 1983, but in March 1983 the National Union of Seamen (NUS) refused to participate in the dumping programme. Because of action by the NUS and other trade unions, NIREX was unable to carry out the planned sea dump. In September 1983, the NUS policy of 'blacking' the sea dumping of solid radioactive waste was adopted by the Trades Union Congress (TUC).

In October 1983 the U.K. government abandoned the 1983 dumping programme, and no radioactive waste was dumped by any country in the Atlantic. The radioactive waste packaged for sea dumping remained in a dry storage building at Harwell. In December 1983 the U.K. government announced that, until a review of the safety of sea dumping had been completed (the Holliday report), further sea dumping would not be authorised. The U.K. government was now acting in advance of receiving authoritative advice, the very reason it had given for opposing the LDC

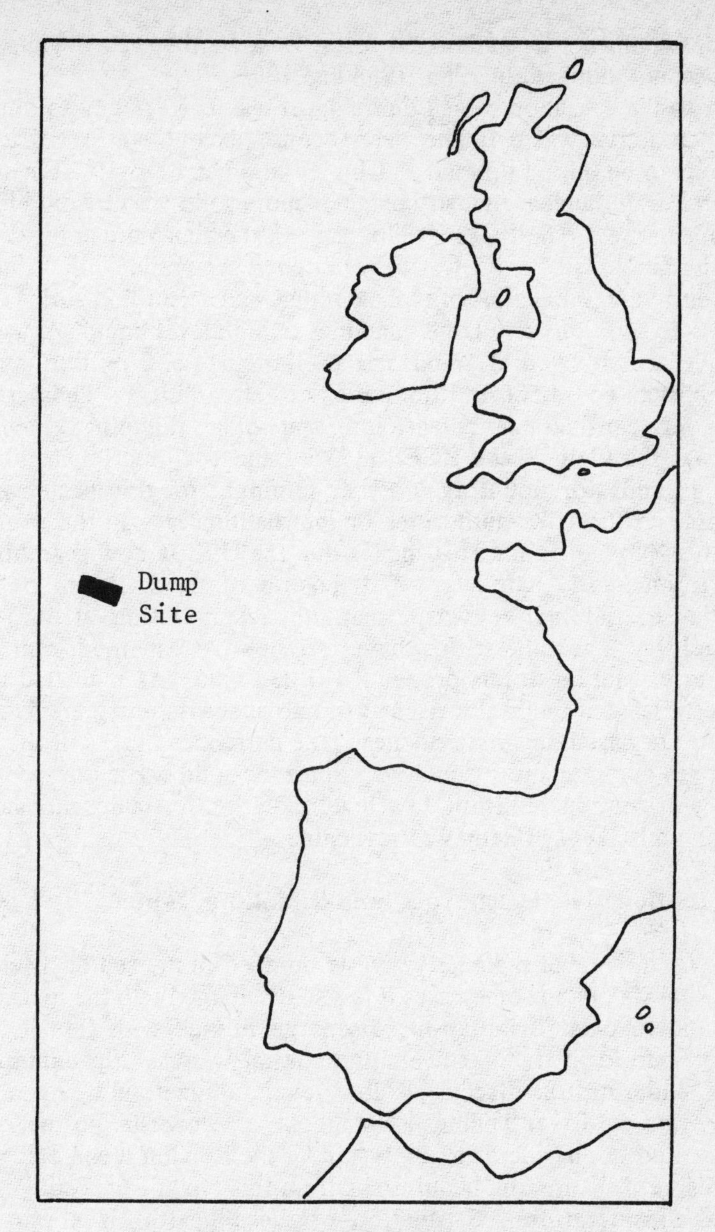

Figure 10.1 The North Atlantic Dump Site

moritorium on sea dumping! Early in 1984 the European Parliament resolved that sea dumping, including liquid waste discharged via the Windscale (Sellafield) pipeline, should not take place. In December 1984, on receipt of the Holliday report, the U.K. government decided that sea dumping would not be authorised for at least another year. Thus, solid waste was not dumped in 1983, 1984 and 1985.

Prior to the cessation of sea dumping, it was U.K. policy to dump all solid radioactive waste in the deep ocean (except waste sufficiently innocuous to be buried on land at Ulnes Walton and Drigg). This meant that as the U.K. nuclear industry expands more solid waste would have to be dumped at sea. The RCEP (1976, page 139) estimated that by 1985 the following amounts of solid waste would have accumulated at Windscale (Sellafield): 210,000 curies of alpha emitters and 15 million curies of beta emitters. In addition, very large quantities of high level liquid waste would also have accumulated at Windscale (Sellafield) i.e. 2 million curies of alpha emitters and 4,500 million curies of beta emitters. These figures imply a very considerable increase in the rate of sea dumping by the U.K., and it was the view of the RCEP (1976, page 142) that 'there are very strong grounds for doubting whether dumping on this scale will be acceptable to the U.K. authorities or internationally.' In the event, the RCEP was proved right. This means that the U.K. is now searching for an environmentally safe way of disposing of the increasingly large quantities of radioactive waste that are accumulating at Windscale (Sellafield). The authorities have effectively assumed that some satisfactory solution to this problem will be found. As indicated above, whilst estimates of the health risks to people of sea dumping are available, these are not based on firm evidence. The current estimates of the health risks of sea dumping are speculative. Since the solid waste dumped in the oceans will continue emitting low level radiation for many thousands of years, any mistake is effectively irreversible.

10.4 Radioactive Discharges and Drinking Water

Not only is low level radioactive waste dumped in the sea or discharged into coastal waters, in some cases it is also discharged into the local rivers. Almost all nuclear installations discharge liquids with low levels of radiation into local rivers, sewers and coastal waters. For example, the AWRE Aldermaston discharges low level radioactive waste down a pipeline into the River Thames at Pangbourne. Downstream the Thames water is used to provide drinking water for the London area. Because of its widespread nature, the health risks from this type of exposure deserve serious consideration. Estimation of these health risks requires a knowledge of the extent to which ingested radioactive material is absorbed through the gut.

Larsen et al (1978) have argued that the health risks from swallowing plutonium dissolved in drinking water have been massively underestimated. Previous estimates have been largely based on laboratory experiments using distilled water. They present results showing that the chlorine in our drinking water oxidises the plutonium, increasing human absorption by 1,000 times. This implies that the previous estimates of the gut uptake factors were something like 1,000 times too low.

In 1979 the ICRP produced revised gut absorption factors for workers, based very largely on studies of adult animals (rats, guinea pigs, hamsters, dogs, mice, rabbits and pigs), Harrison (1981). The old and the new ICRP gut uptake factors are set out in table 10.2. This table shows that all of the ICRP estimates were increased by at least 100%, and in some cases by as much as 100 times. In the animal studies used to make these estimates high doses were introduced directly into the animal's stomachs. The results of the animal studies are not necessarily representative of uptake factor for humans ingesting low doses. The NRPB propose that for adult members of the public exposed to small amounts of plutonium the uptake factor is five times larger than for workers, and almost seventeen times higher than previously estimated, Harrison (1981). For newborn children gut absorption for soluble plutonium is much higher than for adults. This implies that the uptake factor for newborn children is estimated to be roughly 1,700 times larger than previously.

Table 10.2 ICRP Gut Uptake Factors for Workers

Actinide	Old ICRP Factor	New ICRP Factor	Increase
Neptunium	0.01%	1.00%	100 times
Protactinium	0.01%	0.10%	10 times
Americium	0.01%	0.05%	5 times
Curium	0.01%	0.05%	5 times
Uranium (soluable)	1.00%	5.00%	5 times
Plutonium (soluable)	0.003%	0.01%	3 times
Thorium	0.01%	0.02%	2 times

These major revisions to the ICRP factors indicate substantial uncertainty about the risks from drinking radioactive liquids. The latest estimates of the gut uptake factors are still largely based on animal studies and so may not be representative of human factors.

While various nuclear establishments are deliberately discharging radioactive waste into the River Thames, e.g. Aldermaston, Burghfield, Harwell and Amersham International, the only empirical study of the human health effects was only conducted because of an accidental discovery. The Central Veterinary Laboratory at Weybridge, Surrey, were studying the levels of radioactive iodine in the thyroids of sheep in the Windscale (Sellafield) area. As controls, they also measured the quantities of radioactive iodine in sheep at Weybridge. They were surprised to find that both groups of animals had comparable levels of radioactive iodine in their thyroids. The source of the radioactive iodine in the sheep at Weybridge was found to be the drinking water supply, most of which comes from the River Thames.

This discovery prompted a smallscale study into the levels of radioactive iodine in the thyroids of people in these two areas. The thyroids of 14 people living within 22 miles of Windscale (Sellafield) and 39 people from the Weybridge area were examined. As a control, 16 human thyroids from the Dumfries and Galloway area of southwest Scotland were also examined. It was found that all the Weybridge thyroids contained a detectable level of iodine-125, while all of those from Windscale (Sellafield) contained detectable levels of iodine-129. None of the thyroids from Dumfries and Galloway contained any detectable radioactive iodine. The average level of radioactive iodine, measured in becquerels, at Weybridge, was found to be double that for Windscale (Sellafield). While the radiation dose from these contaminated thyroids is well within the permitted levels, in some cases the doses at Weybridge are 100 times higher than for people in other parts of the country. It appears that the source of this radioactivity is the drinking water in the Thames valley area.

Milne (1985) revealed that the Radiochemical Inspectorate was concerned about two companies in west London that are discharging tritium, a highly radioactive gas with a half life of 12.26 years that emits beta particles, into the air. The Inspectorate was worried that, on calm days, the tritium could fall with rain on local gardens. Therefore, the Inspectorate asked the Thames Water Authority to take consignments of tritium-contaminated water, and to mix this contaminated water with effluent from the Mogden sewage works. This effluent is poured into the River Thames. Thus, in order to rule out the possibility of contaminated water falling on people's gardens, it was proposed to put the tritium *directly* into the water system. The Thames Water Authority turned down this request by a government department that the Thames should be deliberately contaminated with radioactive material.

10.5 Uranium Tailings

After being mined, uranium ore is crushed and ground and the uranium is chemically separated. The residue contains the radioactive materials thorium and radium as well as about 4% of the uranium in the uranium ore that has not been extracted. This residue is dumped in tailing piles. Whilst it would reduce radioactive emissions, most of these piles of uranium tailings have not been covered with earth. Pohl (1976) discusses the escape of the radioactive gas, radon-222, from these tailings piles into the atmosphere. The studies of uranium miners, such as those considered in chapter 8, show that radon-222 causes cancer and so these tailings piles will also cause cancer. Pohl points out that an EPA study estimated that over a period of 100 years the release of radon gas from tailings will cause approximately 200 cases of lung cancer worldwide. However, this EPA study grossly understates the total number of lung cancers that will be caused by these uranium tailings because it considers only the first one hundred years. These tailings contain thorium-230 which has a half life of 76,000 years, and this thorium will continue to give off radon gas for many thousands of years. Pohl estimates that the total number of cases of lung cancer caused by the release of radon gas from tailings will be not 200, but 63,000. This dramatically higher figure is because Pohl considers the total effect of these tailings over all future years. In view of this Pohl suggests that, instead of being dumped in piles, tailings should be sealed in abandoned deep mines. Pohl's analysis was largely confirmed by Cohen (1982) when he estimated the risks to be three quarters of Pohl's figure. Grossman (1980, page 5) presents a copy of an NRC memo from Walter Jordan to James Yore, Chairman of the Atomic Safety and Licensing Board Panel in September 1977. After referring to the work of Pohl, Jordan points out that the estimated number of deaths of members of the public from uranium tailings in an NRC report was 'grossly in error', the figures should be 100,000 times higher!

Comey (1975) extended Pohl's analysis to calculate the number of deaths from lung cancer that would occur if the planned expansion of the U.S. nuclear industry to the year 2000 actually took place. Comey calculated that the thorium-230 in the tailings from 30 years operation of these plants plus the tailings already existing in the year 2000 would kill over 11 million people worldwide. Half of these deaths would occur in the next 80,000 years with roughly 3 million deaths in the eastern United States in the next 80,000 years. If allowance is made for growth in the world population, the number of worldwide deaths from lung cancer may be over 25 million. The analysis by Pohl and Comey shows that, if the long term effects of the nuclear industry are considered, the health effects can be massive and that more attention should be given to uranium tailings.

Caldicott (1980, page 27) reports that hundreds of acres of tailings from uranium mining were dumped at Grand Junction, Colorado. In the mid-1960's local construction firms used the tailings to build houses. Rotblat (1982) mentions that inside some of these buildings the levels of radon gas were 1,000 times higher than the natural background level. The use of uranium tailings for construction purposes has now been stopped. Gofman (1983, page 272) quotes a 1979 study of Sweden which demonstrated a significant association between lung cancer and living in homes with increased levels of radon and its daughter products. In 1970 a local Colorado paediatrician noticed an increase in cleft palates, cleft lips and other congential defects among newborn babies in the area. Further investigations found that all these children with congenital defects had been born to parents living in homes built from tailings and that these homes had high levels of radiation. Researchers at the Medical Centre of the University of Colorado then obtained funds from the EPA to study the problem. However, a year later the EPA cut off the funds and the research ceased.

In the Four Corners area of New Mexico, where there are considerable amounts of uranium tailings, Navajo Indian children under the age of 15 have 17 times the expected rate of reproductive organ cancers, 5 times the national rate of bone cancer, 1.7 times the national lymphoma rate and 1.6 times the national rate of cancer of the brain and central nervous system, Bertell (1985, page 69).

10.6 Nuclear Accident in the USSR

What appears to have been the most serious nuclear accident to date, even more serious than Chernobyl, occurred in the Russian Urals in the winter of 1957-58. It is thought there was an explosion involving the radioactive wastes from nearby reactors and plutonium separation plants. It is possible that radioactive waste was dumped into trenches and (as at Hanford in the U.S.A.) concentrations of radioactive material built up at the bottom of these trenches. Water soaking into the soil could have been rapidly heated by the radioactive material and turned into steam. The pressure of this steam might then have produced an explosion with the radioactive material being blown into the atmosphere. An alternative explanation is that liquid radioactive waste was stored in tanks that were insufficiently cooled. The build-up of heat could have led to an explosion. From published studies of wild deer, Medvedev (1980, page 63), estimates the area contaminated with radiation is at least fifty to one hundred square kilometers. Later, on page 143, Medvedev estimates that hundreds of square miles were contaminated by radiation. A study that is reported in the New Scientist (1980) from the U.S. Oak Ridge National Laboratory by J. Trabalka, L. Eyman and S. Auerbach confirms that the accident directly affected an area of 1,000 square kilometers just east of Kyshtym and north of Cheliabinsk. A comparison of maps before and after the accident reveals that the names of over 30 small communities have been deleted from later maps. Medvedev (1980, page 166) reports a local witness as stating that all the hospitals in the area were full of the victims of radioactive contamination and that the total number of patients was in the thousands, of whom most died. These numbers exclude those Russians who subsequently died of radiation induced cancers. Indeed such deaths will continue for another twenty years. This accident reveals the potentially lethal nature of radioactive waste.

10.7 Cancer in Smokers and Alpha Radioactivity

Low concentrations of polonium-210 and lead-210 have been found in tobacco and cigarette smoke, and the importance and implications of this have been discussed by Martell (1974 and 1975). It is thought that the source of this radioactivity is the heavy treatment of the soil in which tobacco plants are grown, with phosphate fertilizers. These fertilizers contain radium-226 and this produces radon-222 which is present at high concentrations in the soil gas and in the surface air under the vegetation canopy provided by the field of growing tobacco. Some of the radioactive decay products of radon-222 then become attached to the tobacco leaves.

The biological significance of the alpha emitters, polonium-210 and lead-210, in tobacco smoke has been disregarded because the quantities involved are roughly equal to the level of alpha emitters in the lungs of non-smokers. However, there is an important difference between these two sources of radiation exposure. The natural alpha activity in the human body is in chemical forms which are soluble in the blood and other body fluids and there is an approximnate balance between daily intake and elimination. In contrast the polonium-210 and lead-210 from smoking are in an insoluble form as smoke particles or particle clusters which tend to accumulate over time. Thus the pattern of irradiation of lung cancer cells is vastly different for insoluble alpha emitters compared with that of soluble alpha emitters. The small volume of tissue around each insoluble smoke particle or particle cluster is subjected to alpha radiation rates between 100 and 10,000 times the natural level. Martell (1975) argues this means that alpha emitters ingested due to cigarette smoking carry a much greater cancer risk than do natural alpha emitters. He postulates that the mechanism by which cigarette smoking causes lung cancer is due to the alpha emitters in the tobacco smoke. Hence the strong epidemiological evidence on the link between cigarette smoking and lung cancer can be viewed as evidence of the effects on health of inhaling insoluble particles of moderate to low alpha activity.

Martell's exposition makes use of the concept of 'hot particles'. A hot particle has been defined in Lovins et al (1975) as an alpha-emitting particle in the deep respiratory tissue of such activity as to expose the surrounding lung tissue to a radaition dose of at least 1,000 rem (10,000 mSv) per year. The hot particle argument appears to imply a convex dose-response curve and so to be in conflict with the evidence that small doses of radiation are relatively more dangerous. There does not appear to be any direct evidence about the relative dangers of hot particles, and further empirical and theoretical work on this topic is needed.

10.8 Natural Radiation

In chapter 5 it was stated that each person is exposed to 0.2 rems (2 mSv) per year due to naturally occurring radiation. It has been argued previously that any dose of man-made radiation, no matter how small, will produce an adverse health effect. This suggests that any dose of radiation, whether natural or artificial, will generate cancers etc. If everyone were subject to exactly the same level of natural radiation such a hypothesis would be impossible to test. However, there are small variations in the doses of cosmic and geological radiation throughout the world, making it possible to test the hypothesis. As the variations in natural radiation are very small, large populations are necessary to get results showing a statistically significant effect. Doses of cosmic radiation vary with the latitude and altitude whilst geological radiation varies with the local rock type.

10.8.1 Cosmic Radiation

Cosmic rays are absorbed by the atmosphere, so exposure to such rays is greater at higher altitudes. The magnetic field of the earth influences the path of incoming charged cosmic nuclear particles so that cosmic radiation reaching the surface of the earth tends to be concentrated in areas of low horizontal geomagnetic flux (HGF). For the U.S. the 'contour' lines of equal levels of HGF run east-west and so correspond roughly with the lines of latitude. The level of HGF increases towards the north. Archer (1978) found that for a considerable number of types of cancer which are not known to be caused primarily by some other factor (e.g. tobacco, sunlight, alchohol), incidence in the U.S. roughly increased towards the north pole. Hence cancer rates increased with increasing cosmic radiation. Archer then extended his analysis to the whole world (omitting substantial black populations and areas with large immigrations since 1900). Again he found a positive relationship between cosmic radiation and cancer incidence.

Archer (1979) also examined the correlation between various types of cancer for the U.S. states and cosmic radiation (estimated using HGF). He found the following types of cancer to be correlated with the cosmic radiation dose: kidney, breast, multiple myeloma, female sex organs (except cervix), lymphoma, stomach, testes, prostate, connective tissue, leukaemia, Hodgkin's disease, colon, rectum, thyroid, brain and nervous system, in order of decreasing correlation. Since cancer of the kidney appeared to conform best, Archer conducted a worldwide analysis of the relationship between cosmic radiation and cancer of the kidney. He excluded data from countries with large black populations, a rapid increase in the population or a small population. He found a strong correlation between cancer of the kidney and doses of cosmic radiation. Archer

(1978) concluded that since increases in cosmic radiation led to increases in the cancer rate, studies of the carcinogenic effects of natural radiation should allow for differences in the level of cosmic radiation by holding the level of HGF constant.

Archer (1978) then investigated the effect of altitude on cancer rates in the U.S. by comparing Colorado state with ten other states having a similar level of HGF. Since the bulk of the population in Colorado live at above 1,500 metres they have a thinner blanket of air to protect them from cosmic radiation. He found that for the period 1950 to 1969 death rates in Colorado from cancer of the kidney, breast, stomach, bladder, pancreas, lung and lymphoma were all significantly higher than in the other ten states, i.e. that the area exposed to higher cosmic radiation had higher cancer rates. Of course, part of this effect may be due to the presence of the Rocky Flats plant in Colorado state.

The connection between deaths due to congential malformations and cosmic radiation was studied by Wesley (1960). He used the level of HGF to estimate the quantities of cosmic radiation throughout the world. For seventy countries, covering both the north and south hemispheres, he found a positive correlation between the death rate from congential malformations and the cosmic radiation dose. This correlation was highly statistically significant (0.001%). For 49 U.S. states he found that the positive correlation was again highly statistically significant (0.1% for whites and 0.05% for negros). Wesley concluded that roughly 96% of the variations in the death rate from congenital malformations between countries can be explained by variations in cosmic radiation.

Archer (1979) considered the rates of anencephalus, a congenital abnormality in which the brain is absent, for thirty Canadian cities. Since there is a time lag between exposure to additional cosmic radiation and an increase in anencephalus, migrants will have rates of anencephalus that reflect the cosmic radiation in the cities they have left and not the city where they now live. To allow for this effect Archer divided the cities into those with large and small proportions of migrants. For the latter group of cities he found a positive correlation between anencephalus and cosmic radiation (using HGF to estimate the cosmic radiation dose). The rate of anencephalus in cities with a large proportion of migrants was not correlated with the cosmic radiation dose. This is explained by migrants generally coming from the south with a lower exposure to cosmic radiation and so having a lower rate of anencephalus.

10.8.2 Geological Radiation

In a number of parts of the world the local rocks contain radioactive material which give doses of low level radiation to the local population. There have been studies of the health effects of such geological radiation in

India, China, Brazil, Japan, Austria, Finland and Scotland. These studies, which are summarised below, show that very low levels of radiation can produce adverse health effects.

A coastal area of the south Indian state of Kerala has high levels of natural radiation of 1.5 rem to 3 rem (15 mSv to 30 mSv) per year due to the presence of thorium containing monazite material in the soil. Kochupillai et al (1976) carried out a house-to-house survey of developmental abnormalities for 12,918 people. They used as their control 5,938 people living in a coastal area to the north that is similar in all respects, except that it does not have a high level of natural radiation, see figure 10.2. They found the rate of severe mental retardation due to genetic causes was over four times higher in the area of high geological radiation (2.76 per 1000 as against 0.67 per 1000). They conclude that radiation induced genetic damage is the most likely explanation.

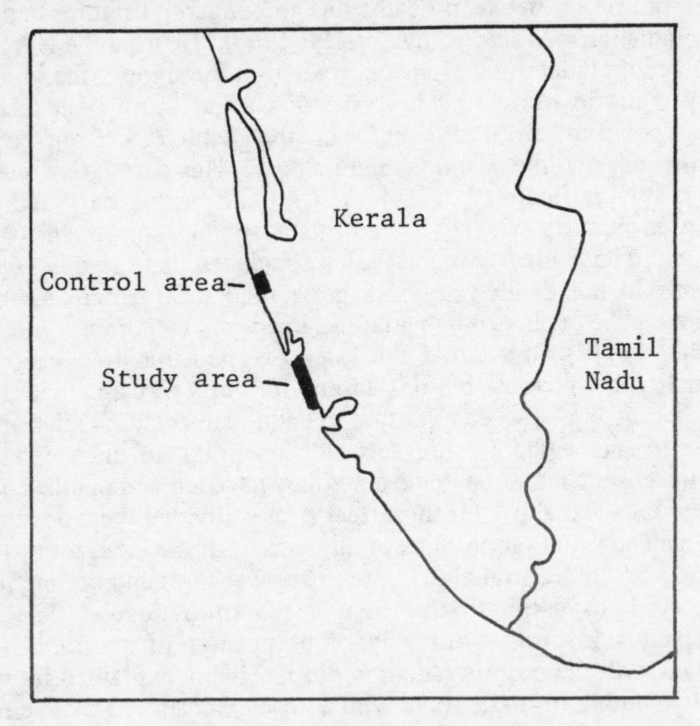

Figure 10.2 Kerala, South India

1 = Control area, 2 = Study area

Barcinski et al (1975) have studied the effects of low level geological radiation on chromosome damage in a small coastal village in Brazil. The village is built upon monazite sand containing 6% thorium and 0.3%

uranium impurities. The average natural radiation dose was determined by issuing 500 local residents with dosimeters to be hung round the neck. The average dose was found to be 0.64 rads (6.4 mGy) per year which is about six times the normal level of natural radiation. Blood samples were taken from 200 people who had lived in the area for at least eight years and had not recently been exposed to medical radiation. As a control, blood samples were also taken from 147 people from another village that is similar except that it has normal levels of natural radiation. The amount of chromosome damage for the two villages was compared and found to be greater for the first village at the 5% level of statistical significance. Therefore, living on monazite sand and so being subject to a very small increase in natural radiation produces a significant increase in chromosome damage. The authors conclude that this chromosome damage may not be due to the increase in external radiation as measured by the dosimeters but to internal radiation which it was not possible to measure. The thorium in the monazite decays to radon gas (radon-220) and this gas decays to lead-212. Hence local residents receive a dose of radiation to their lungs and subsequently to their blood, and it is this internal dose which may be causing the chromosome damage.

In some parts of Yangjiang County, Guangdong Province, China, which is on the coast just to the west of Hong Kong, monazite has raised level of geological radiation to about three times that in the neighbouring areas. A study by the High Background Radiation Research Group (1980) has examined 3,504 people in the study area and 3,170 people in two nearby areas, which were used as a control, see figure 10.3. The average external dose from background radiation was 1.96 rems (19.6 mSv) in the study area and 0.72 rems (7.2 mSv) in the control areas. It was found that the incidence of Down's syndrome in the study area (1.71 per thousand) was significantly higher than in the control areas (zero).

Badgastein, Austria is a health spa built around a hot spring. The spring water contains radon, which is an alpha emitter. The result is that radon-222 is released into the air from the water. Radon gas is also discharged from the ground throughout the Badgastein area. Pohl-Ruling et al (1976 and 1979) took 180 blood samples from 122 people living and working in the area and analysed 30,769 cells for cell aberrations. It is ironic that the highest doses of radiation were found in the main treatment area of the spa. Using estimates of the radiation dose to the blood, Pohl-Ruling et al plotted the combined alpha and gamma dose against the number of chromosome aberrations per 100 cells. The resulting dose response curve is clearly and strongly concave rising very steeply at first up to 0.2 to 0.4 rad (2 to 4 mGy) and then becoming flat. The total radiation dose to the blood was then disaggregated into its alpha and gamma components and it was found that, at annual doses of 0.1 to 0.34 rads (1 to 3.4 mGy) of gamma radiation and 0.001 to 1.6 rads (0.01 to 16 mGy) of alpha radiation, a dose

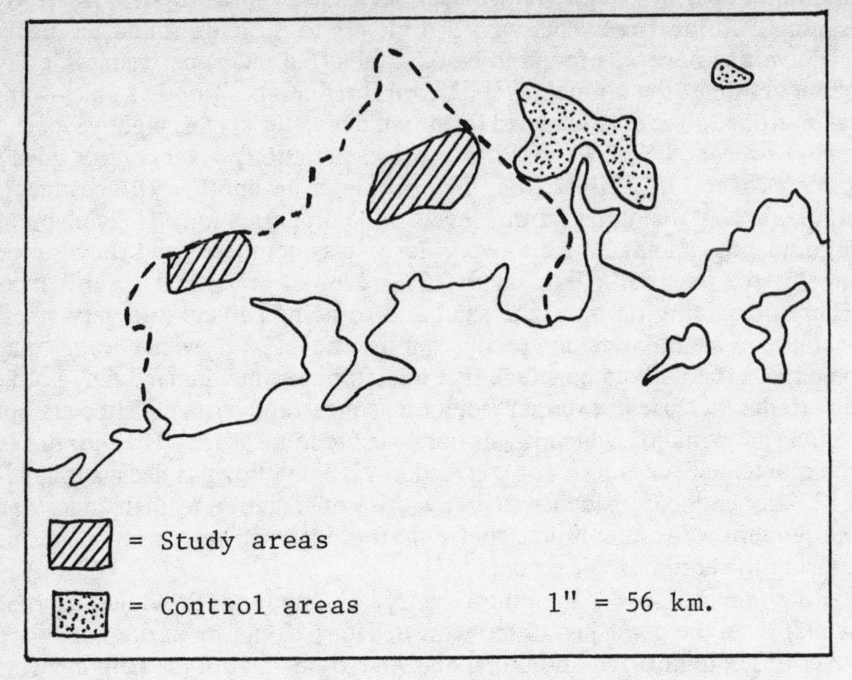

Figure 10.3 Guangdong Province, China

response effect that is positively sloped can be demonstrated. This shows that very small variations in natural radiation can have a demonstrable effect on levels of chromosome aberrations. It is the firm opinion of Gofman (1983, page 271) 'that the only effect a patient could expect from such "therapy" would be greatly increased risk of lung cancer death'.

Uzunov et al (1981) have measured the radiation dose in WLM per year to various groups of employees at the nineteen spas in Badgastein and also the dose to members of the public living in the town. Since no epidemiological study of lung cancer has been conducted at Badgastein, Uzunov et al multiplied the doses by an UNSCEAR risk factor for WLM estimated using uranium miners. This gave the estimated increase in the incidence of lung cancer per million adults per year, for each of the exposure groups, and these are set out in table 10.3. Uzunov et al also calculated the doses at the Bulgarian spa of Momin Prohod and the implied increase in the incidence in adult lung cancers from exposure to the radon gas and its daughters. These figures are also contained in table 10.3, which shows that radon spas can cause a considerable increase in lung cancer, particularly for those occupationally exposed.

Table 10.3 Estimated Exposure and Increase in Lung Cancer Incidence

Exposure Group	Average Exposure in WLM per year	Estimated Lung Cancers per year per million adults
Public Exposure - Badgastein		
Inhabitants living in areas with lower levels of radiation in the air	0.3	105
Inhabitants living close to the spas with higher levels of radiation in the air	1.7	675
Occupational Exposure - Badgastein		
Bath attendants	3.0	1,225
Thermal gallery administrators	3.7	1,575
Thermal gallery doctors & inspectors	6.9	2,880
Thermal gallery miners & trainleaders	35.0	12,000
Patient Exposure - Momin Prohod		
Thermal baths	0.1	45
Inhalation treatment	0.2	72
Public Exposure - Momin Prohod		
Domestic use of hot spa water	0.4	190
Occupational Exposure - Momin Prohod		
Bath attendants	6.0	1,950

In some areas of Finland there are high concentrations of radioactivity in the local water supply. Stenstrand et al (1978) studied chromosome aberrations among 18 persons living in five different houses where radon-rich household water was used, and 9 controls who had been analysed as part of a study of nuclear power plant workers. The rate of chromosome aberrations in those exposed to the radon-rich water was 100% higher than for the controls. Since some 100,000 people are exposed to radon-rich water in Finland the authors call for an epidemilogical study of lung cancer among these people.

In parts of the U.S. states of Iowa and Illinois, the drinking water has a high level of radium-226. Petersen et al (1966) studied 111 communities (72 in Illinois and 39 in Iowa) where the drinking water had over 3 pica curies per litre (average 4.7 pica curies) due to the radium-226, and also 111 control communities of similar size and age in these states where the level of radium-226 in the drinking water was less than 1 pica curie per litre. It should be noted that in 1962 the U.S. Public Health Service set a maximum limit of 3 pica curies per litre for radium-226. The authors then compared the death rate from bone cancer from 1950 to 1962 for these two groups. They found a 24% increase for those communities exposed to high levels of radium-226 in their drinking water, which was statistically significant at the 8% level.

Another study of the health effects of radioactivity in Iowa's drinking water was conducted by Bean et al (1982). They studied communities in Iowa with a population between 1,000 and 10,000 people whose drinking water came solely from deep wells and was not softened. These 28 towns were divided into three groups according to the radium-226 content of their drinking water (a) under 2 pica curies per litre, (b) 2 to 4.9 pica curies per litre and (c) over 5 pica curies per litre. The cancer incidence rates for these three groups for the years 1969 to 1978 (excluding 1972) were compared. The male lung cancer rate was 68% higher for the high dose group than the lower dose group, and this increase was statistically significant at the 0.2% level. After allowing for differences in median income, occupation and water fluoridation, the increase in male lung cancer was still significant at the 2.8% level.

In the Cote-du-Nord area of northern Brittany, some villages are located on uranium deposits and so the inhabitants are subject to increased levels of background radiation, see figure 10.4. Neighbouring villages are not subject to such an increase in background radiation. Pincet et al (1975) identified two areas, each containing three villages which are located on the uranium deposits, and two nearby control areas which are not built on uranium deposits. The population of the study areas was 7,000 and that of the control areas was 8,000. Using the age adjusted rate of cancer deaths in Brittany in 1968, they calculated the expected number of cancer deaths for each area for the five year period 1969-73. This was then compared with the observed number of deaths for each of the four areas. For the areas containing uranium deposits the cancer death rate was 43% higher than expected, whilst for the control areas the cancer rate was 7% lower than expected. The 43% excess for the high background radiation areas was statistically significant, whilst the 7% deficiency in the control areas was not. Pincet et al also compared the observed number of deaths from stomach cancer in the areas of high background radiation and the control areas, with the number expected on the basis of the rate for France in 1971. The rate of stomach cancer deaths for the study areas was 132% higher than expected whilst for the control areas it was only 27% higher. The 132% increase in the areas of high background radiation was statistically significant, whilst the 27% excess in the control areas was not.

Court Brown et al (1960) calculated the average levels of natural radiation (cosmic and geological) for ten areas covering Scotland. The cosmic radiation was then subtracted to give the geological radiation doses. A comparison was made between Aberdeen, which is built on granite, and Dundee, which is located on sedimentary rocks. In each town most houses are built of the local stone. It was found that the annual dose to each

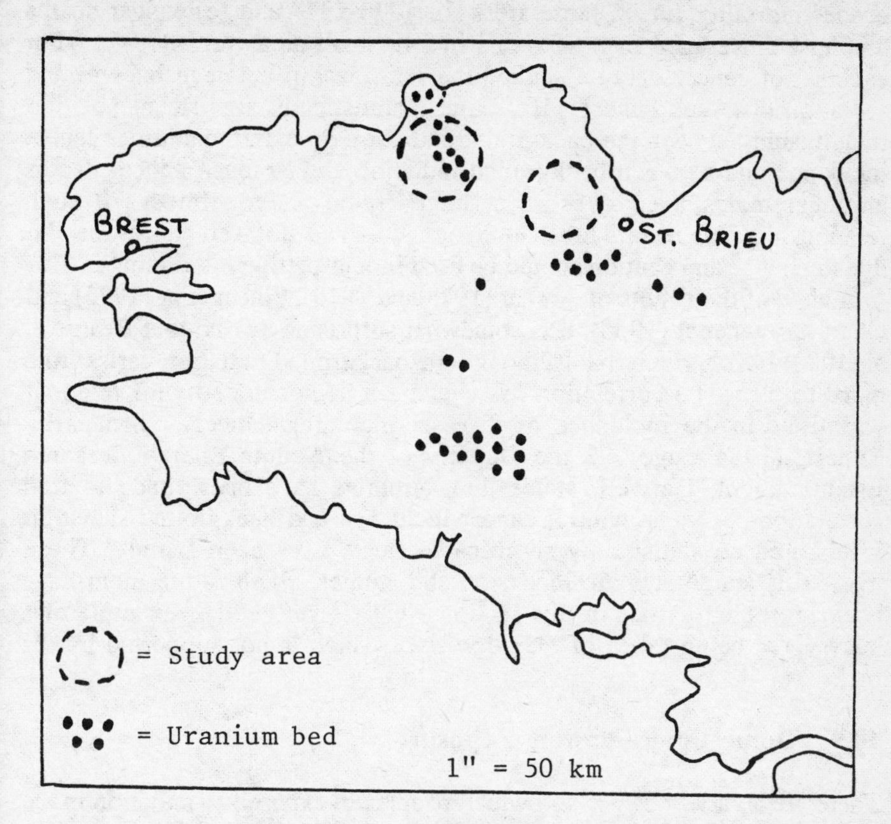

Figure 10.4 Brittany

person's bone marrow from geological radiation was 0.043 rad (0.43 mGy) in Dundee and 0.058 rad (0.58 mGy) in Aberdeen, i.e. a difference of 0.015 rad (0.15 mGy). The authors then compared the standardized death rates from acute leukaemia and chronic myeloid leukaemia for those aged 15 and over for these two towns for the period 1939 to 1956. They found that the death rates were 50% to 60% higher in Aberdeen than in Dundee, whilst the level of geological radiation was 35% higher in Aberdeen. Court Brown et al (1960) conclude that the higher levels of natural radiation in Aberdeen were not responsible for the higher leukaemia rates, because to do otherwise would be to accept that very low levels of radiation were over 100 times more dangerous than previous estimates suggested!

The health effects of background radiation in Japan have been studied by Ujeno (1983). He used official data to calculate the background dose for each area of Japan. He then analysed three sets of cancer data; (a) age adjusted cancer incidence in 13 large areas for 1975, (b) standardized cancer mortality for 46 large areas for 1973-1977 and (c) cancer deaths amongst those aged over 40 years in 649 small areas for 1969-71. For each set of cancer data he studied the relationship between background radiation dose and cancer. He found a statistically significant positive relationship between the background radiation dose and male liver cancer incidence and between background radiation and stomach cancer deaths amongst males aged over 40. Since Ujeno tested almost 70 such relationships, his results for male liver cancer and stomach cancer may be due to chance and caution should be used in interpreting his findings.

In view of the results of Archer (1978 and 1979), Pincet et al (1975) and Court Brown et al (1960), it is somewhat surprising to find that Beattie et al (1980) have written that "although the background radiation varies from place to place, no correlation has ever been demonstrated with regional variations in the incidence of disease, including cancer". Similarly, Roberts (1984, page 20), the Director of the Atomic Energy Research Establishment, Harwell, states that 'attempts have been made to find correlations between natural cancer incidence and background radiation levels, and no statistically significant effects have been found'. These reassuring statements are incorrect, and studies which refute them have been presented earlier in this section. This is yet another example of a 'rosy view' being taken of radiation risks which is not supported by the facts.

10.9 Domestic Radiation Exposure

There are a variety of ways in which people are exposed to radiation in the normal course of their daily lives. Some of these sources of radiation are unexpected, and people may be unaware that they are being exposed to radiation. But like all radiation exposure, this exposure can prove fatal. Gofman (1983, page 348) reports that some optical lenses contain up to 30% by weight of uranium or thorium and that these lenses can deliver appreciable doses to the lens of the wearer's eyes. For example, for glasses containing 18% thorium by weight the absorbed dose was found to be one millirad (0.001 mGy) per hour. If a person wore these glasses for 15 hours a day the annual dose would be 5.5 rad (55 mGy) per year.

Roughly one in nine adults has artificial porcelain teeth. In 1942 it was discovered that uranium causes artificial porcelain teeth to fluoresce, and uranium compounds were used to make false teeth which simulated the fluorescense of natural teeth. Hence, most people with dentures had uranium inside their bodies. An NRPB report by O'Riordan et al (1974) estimated that this resulted in an annual dose equivalent of 2.7 rem (27 mSv) to each person with these dentures. This dose is largely independent of the number of false teeth per person. In 1974 there were no explicit controls in the UK on such radiation exposure, and the doses probably exceeded the ICRP recommended dose limit for the general organs of members of the public. This situation appears to have occurred because the problem of radioactive teeth was overlooked, and the NRPB report recommended the use of non-radioactive fluorescers.

There are many clocks, watches, etc. with luminous dials. They glow in the dark because they have been painted with a mixture of a chemical that glows when exposed to radiation (a scintillator) and some radioactive material, such as radium. The substantial health risks to radium dial painters were described in chapter 8. Gofman (1983, page 346) has estimated that in the U.S. each year 23 people will die from cancers caused by their use of luminous timepieces.

The cathode ray tubes of television sets produce X-rays and so TV sets are thought by the public to be dangerous sources of radioactivity. However, in fact TV sets emit very little radiation. The average dose from a colour TV set to a viewer some two metres in front of the screen has been estimated by Gofman (1983, page 350). He calculated that, for someone who watched TV for 24 hours per day, the dose is only 0.0005 rad (0.005 mGy) per year.

Many rocks contain small quantities of radium-226, which decays to produce the inert radioactive gas, radon-222. Thus building materials can contain small quantities of radium, and so radon gas is continually released into the building. The NRPB has studied the exposure to radon in UK houses, Wrixon (1986). A survey found that the average radon dose in Cornwall houses was over fifteen times higher than the average for the UK. The dose in a very small number of Cornish houses was found to be over one hundred times the national average. One Cornish home was found by the NRPB where the inhabitants received an annual dose of 32 rem (320 mSv), Lean(1986). Using the ICRP risk factors, Wrixon calculated that radiation from radon in homes is causing 400 deaths per year in the UK. Six scientists at the Lawrence Berkeley Laboratories of the University of California have considered the health effects of the decreased ventilation in private homes achieved in the interests of energy conservation. They have estimated that this improved home insulation could cause between 1,000 to 10,000 extra deaths from lung cancer each year in the UK, Dickson (1978).

10.10 Some Nuclear Accidents

Studies have found that the nuclear accident at Windscale (Sellafield) in 1957 killed over twenty members of the public, but have been unable to identify the people concerned. This is a common problem where exposure to low levels of radiation is concerned, for the reasons given in chapter 4. However, there have been some nuclear accidents where some or all of those killed can be specifically identified. In chapter 8 some radiation accidents leading to the deaths of identified workers were described. There have also been radiation accidents which have killed particular members of the public, Pochin (1983, pp.99-100).

In Mexico in 1962 a five curie source of cobalt-60 was being used on a building site to check the metal structure for cracks and defects. It was found by a ten year old boy who placed it in a wooden cupboard in his home. He and three other members of his family died within months from the high doses of radiation they received from the cobalt source. During 1963 a ten curie source of cobalt-60 was being used for irradiating seeds in China. It was found by a boy and taken home. He and his brother are estimated to have received doses of 8,000 rem (80,000 mSv) and 4,000 rem (40,000 mSv), and they were killed by this radiation. In Algeria in 1978 a twenty five curie source of iridium-192 used for industrial radiography was taken home and killed one person.

The accidental exposure of the public to radiation is happening all the time. For example, a somewhat unusual accident recently occurred in Mexico. In December 1983 scrapyard workers in Juarez broke open a discarded cancer therapy machine releasing over 6,000 tiny pellets of highly radioactive cobalt-60. Many of these pellets were scooped up with the other scrap metal and melted down to make steel reinforcement rods. These rods have been incorporated into buildings in the USA and Mexico. Other contaminated steel has been used to make table legs. These legs were shipped to many American states for use in restaurants, hotels, shops, etc. The accident was only discovered in January 1984 when a delivery truck carrying some of the radioactive steel rods took the wrong route and triggered radiation alarms when it passed the Los Alamos National Laboratory, Wilkinson (1984). The radioactive table legs were also discovered by another happy accident. A truck carrying some of these radioactive table legs to a Chicago restaurant set off a radiation detection device in a police car near Chicago, Barlett et al (1985, pp.325-326).

10.11 Summary

There is inadequate knowledge of the environmental pathways by which radioactive material released to the environment returns to humans. This is particularly true in the case of the dumping of solid radioactive waste in the oceans. Despite this lack of knowledge of the long term health effects, the expansion of the UK nuclear industry implies an increase in the UK's sea dumping or an expansion of land disposal. Calculations of the health risks from uranium tailings indicate there will be many millions of deaths wordwide from lung cancer in future years. These alarming predictions do not appear to have provoked much response. Indeed, uranium tailings have tragically been used in the construction of houses. The storage of nuclear waste in the USSR has resulted in a major accident killing thousands of people and contaminated an area variously estimated to be from 100 to 1000 square kilometers.

Evidence has been produced that low level radiation is the mechanism by which smoking causes lung cancer. The doses of low level radiation are delivered to the lungs by radioactive particles derived from fertilizers used on the tobacco plants. This implies that all the evidence about the causal link between smoking and lung cancer is evidence of the health effects of low level radiation. Finally, various studies of the health effects of cosmic and geological radiation show that very small variations in natural radiation can cause variations in the rates of cancer, Down's syndrome and chromosome damage. According to Nobel Laureate in Medicine, James D. Watson

> "I fear that when the history of this century is written, the greatest debacle of our nation will be seen not to be our tragic involvement in Southeast Asia but our creation of vast armadas of plutonoium, whose safe containment will represent a major precondition for human survival, not for a few decades or hundreds of years, but for thousands of years more than human civilization has so far existed"

quoted by Grossman (1980, page 82).

11 Cost-benefit analysis of radiation exposure

11.1 Introduction

So far this book has concentrated upon the health risks of radiation exposure. A common response of the authorities to evidence that some practice is causing cancers is that the level of radiation exposure is well within the safety standards i.e. permissible. This moves the subject of debate from the question 'how dangerous is low level radiation?' to the question of 'are the risks acceptable?' This second question, concerning whether the permitted risks are acceptable, necessitates a balancing of the risks against the associated benefits. This chapter discusses the problems and techniques involved in such a cost-benefit approach. It is somewhat surprising to discover that the use of a formal cost-benefit analysis for appraising radiation exposure is a comparatively recent phenomenon. At least in the UK, such an approach has been used only for the past few years. Previously, there does not appear to have been any formal procedure for balancing the costs against the benefits.

11.2 The Optimal Level of Activity

Radiation exposure standards can only be validly set by considering all the costs and benefits to society of the various activities that involve radiation exposure. Welfare economists have shown that, given certain assumptions, a necessary condition for maximizing social welfare is that each activity be expanded or contracted until the marginal benefit received by society is equal to the marginal cost to society i.e. the net marginal social benefit equals zero. The costs and benefits to society are known as social costs and benefits.

Figure 11.1 shows some marginal and total social cost curves and the relationship between them. The lower part of figure 11.1 shows total social costs and benefits, and the upper part contains marginal social costs and benefits. (The private cost curves in figure 11.1 should be disregarded for the moment). At every level of activity each marginal curve gives the amount by which the corresponding total cost or benefit will increase if the level of activity is increased by one unit. For example, the marginal social cost involved in a telephone call is tiny e.g. the small amount of electricity required in making the connection. The total net benefit to society is maximized at the level of activity of OA in figure 11.1 i.e. when marginal social cost equals marginal social benefit. This occurs at point H, where the marginal social cost and benefit curves intersect. The net social benefit is given by the difference between the total social cost and benefit curves. This is maximized at a level of activity of OA, when the total net benefit to society is given by the distance BD.

In constructing figure 11.1 it has been assumed that marginal social cost is constant with respect to the level of activity. For example, a linear dose-response curve implies that the marginal cost of adverse health effects is a constant. To ensure that the socially optimal level of activity is positive, it has been assumed that marginal social benefit decreases as the level of activity increases i.e. that the total social benefit curve is a concave function. Other assumptions could have been used for the shapes of the social cost and benefit curves, but a number of the possible alternatives imply that the socially optimal output is zero e.g.

(a) constant marginal social cost and increasing marginal social benefit.

(b) decreasing marginal social cost and constant marginal social benefit.

(c) decreasing marginal social cost and increasing marginal social benefit.

(d) marginal social cost increasing at a slower rate than marginal social benefit.

(e) marginal social cost decreasing at a faster rate than marginal social benefit.

(f) marginal social cost exceeds marginal social benefit, and both are constant.

Because they imply zero output, these alternative assumptions about the shape of the social cost and benefit curves are not presented in diagrammatic form.

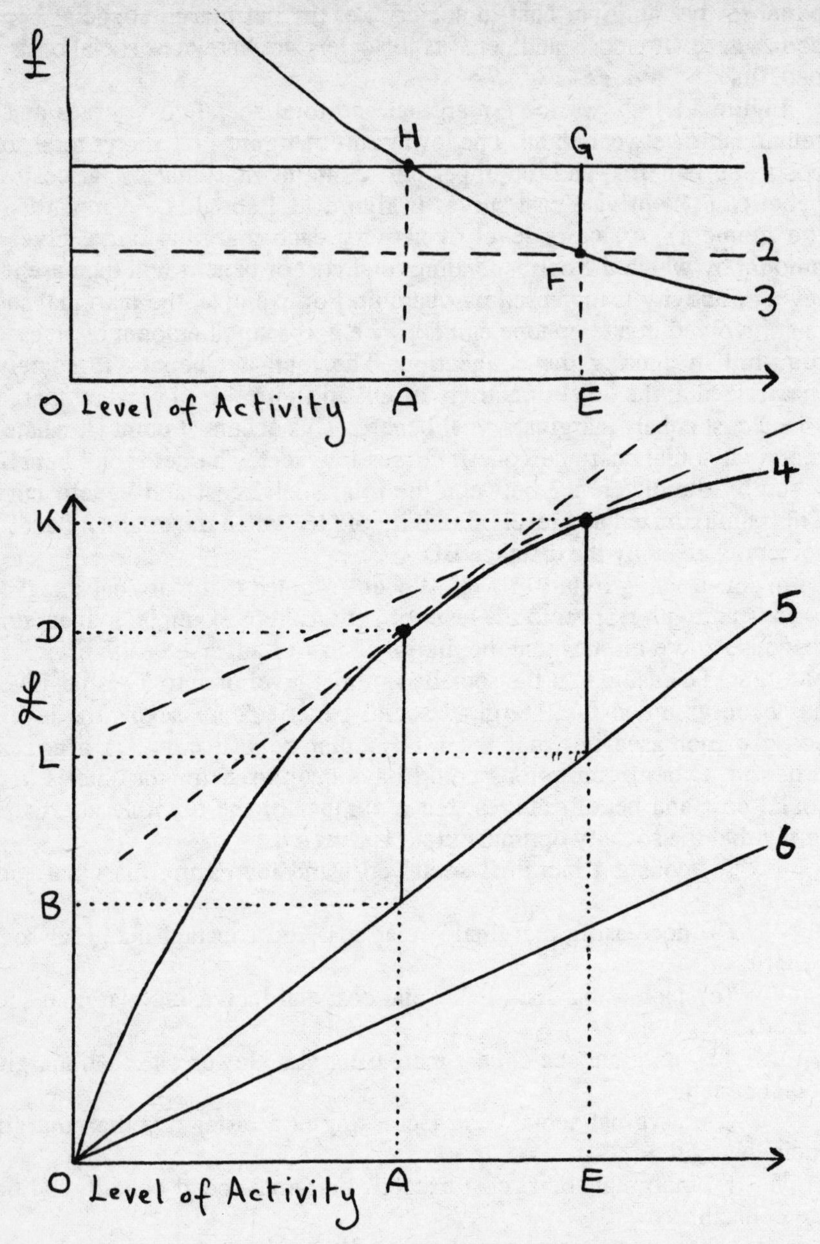

Figure 11.1 The Optimal Level of Activity

1 = Marginal Social Cost, 2 = Marginal Private Cost, 3 = Marginal Benefit, 4 = Total Benefit, 5 = Total Social Cost, 6 = Total Private Cost

11.3 Safety Measures and Social Costs

An important factor in determining the total social costs associated with a particular level of activity is the production technique used e.g. whether an AGR or a Magnox nuclear reactor is used to generate electricity. It may be assumed in constructing a total social cost curve that, for each particular level of activity, the cost minimizing production technique may change as the level of activity alters. So, in order to maximize social welfare, two decisions must be made, (a) the level of activity of the radiation-generating activity, and (b) the production technique to be used. The choice of an overall production technology e.g. type of reactor, will not be considered further. However, it must be remembered that an optimal choice of production technology is implicit in the construction of the curves in figure 11.1 and different points on the social cost and benefit curves may correspond to different technologies.

One important feature of the choice of production technology is the safety measures to be incorporated within the production process, and this aspect of the technology will be considered explicitly. For example, considerable sums of money are spent on the treatment, storage and disposal of the radioactive waste from nuclear reactors. Alternatively, this waste could just be dumped in some nearby field, as are the radioactive tailings from uranium mines. Expenditure on safety measures will reduce the adverse health effects associated with each level of activity. Provided the marginal cost of these adverse health effects exceeds the marginal cost of providing the safety measures, social cost will be reduced by an expansion of safety measures. Thus, for any given level of activity, expenditure on safety measures should be expanded until the marginal reduction in total social cost is zero.

With a suitable reinterpretation of the axes and curves, figure 11.1 can be used to illustrate the problem of selecting the optimal level of safety measures. This is shown in figure 11.2. In this diagram both curves are concerned with costs since the benefit of safety measures is a reduction in adverse health effects. Thus, the optimal quantity of safety measures is OQ i.e. when the marginal reduction in total social costs from an expansion of safety measures is zero. At this level of safety measures, the total reduction in social costs due to the safety measures is given by the distance MN. The social cost curves in figure 11.1 can be thought of as being constructed for a given level of safety measures. Note that, whilst the level of safety measures is being held constant along these curves, the overall technology may vary. Similarly, the social cost curves in figure 11.2 for safety measures are drawn for a particular level of operation of the radiation-generating activity. Hence, the level of activity, the level of safety measures and the overall technology must all be simultaneously optimized.

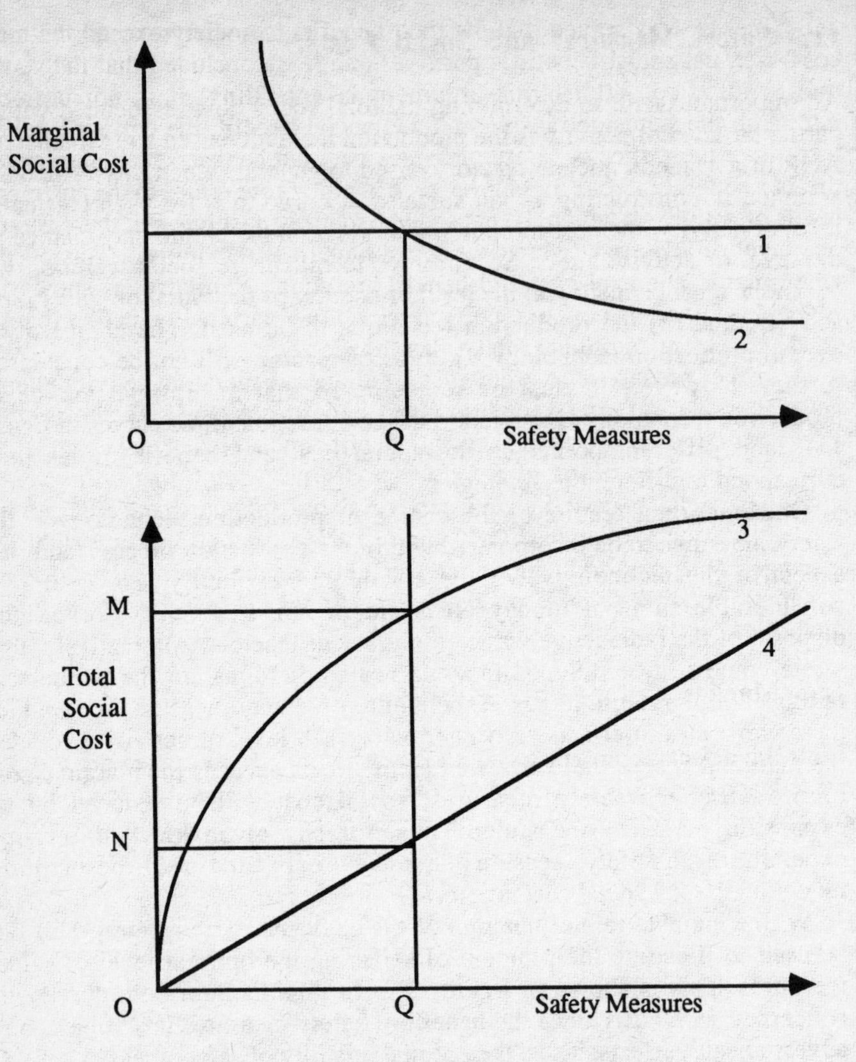

Figure 11.2 Optimal Level of Safety Measures

1 = Marginal Social Cost of the Safety Measures, 2 = Reduction in Marginal Social Cost due to the Safety Measures, 3 = Reduction in Total Social Cost due to the Safety Measures, 4 = Total Social Cost of the Safety Measures

11.4 Marginal and Total Social Costs and Benefits

Often the analysis of some radiation-generating activity is conducted in terms of total social costs and benefits, rather than marginal social costs and benefits. It is argued that the total benefits to society exceed the total costs (i.e. net social benefit is positive) and it is concluded that the status quo is justified. Whilst this argument is appealing, it is not correct. Although the status quo results in a net social benefit, the current level of activity or the choice of safety measures may not be that which maximizes the net social benefit. As can be seen from figure 11.1, there are many levels of activity at which the net social benefit is positive. But there is only one level of activity at which net social benefit is maximized. This occurs when marginal social cost and benefit are equal. Similarly, as shown in figure 11.2, there is only one point at which the marginal reduction in total social cost from additional safety measures is zero.

A further mistake is to argue that a radiation-generating activity should be expanded until total social cost and benefit are equal. If the level of activity is expanded until total social benefit equals total social cost the net benefit to society drops to zero, and marginal social cost will exceed marginal social benefit i.e. the net social benefit can be increased by contracting the level of activity. Hence, it is marginal and not total (or average) social cost and benefit that should be equated. Similarly, the level of safety measures should only be increased until the marginal reduction in total social cost is zero.

11.5 Market Failure and Radiation

If all the effects, both good and bad, of a radiation-generating activity are traded across competitive markets then, given certain assumptions, an economically efficient level of operation for the activity will be set, and no interference by the government in setting exposure standards is indicated. Such costs and benefits which are traded across markets by the producer are called private costs and benefits. But radiation-generating activities do not, indeed cannot, have all their effects traded across markets. For example, pollution of the environment from nuclear weapons testing or the operation of a nuclear power plant is not traded on any market. Such market failure leads to what are called 'negative externalities' i.e. an involuntary reduction in the utility of those people affected by the pollution. Social cost can now be defined as the sum of the private costs and the negative externalities.

Markets in pollution do not exist because of the practical difficulties inherent in their operation. Thus, if people have the right not to be polluted, a would-be polluter must buy the right to pollute every person concerned. This will generate considerable transactions costs e.g. negotiating with thousands or millions of people, possible in foreign countries. In addition, a few people may refuse to sell at any price their right not to be polluted. This will negate the deals reached with willing sellers and block the operation of the pollution creating activity.

If the allocation of property rights is such that the people affected have to buy the right not to be polluted from the firm, there will also be market failure. To buy the pollution rights the public must form a voluntary association to negotiate with the firm and to raise the money with which to buy the right. However, there is a 'free rider' problem in that it is in each individual's interest to let others contribute towards buying the pollution rights. Therefore, if everyone follows his or her own self-interest, no-one will contribute towards purchasing the pollution right.

To some extent, market failure may be corrected by the legal system. People who have been subjected to a negative externality can sue the producer to compensate them for the damages suffered. For example, the Nuclear Installations (Licensing and Insurance) Act 1959 and the Nuclear Installations (Amendment) Act 1965 give people the right to seek compensation for up to thirty years. But, in the view of Justice Parker(1978, page 61),

> 'the apparently beneficial effect of the provisions is, however, likely in the ordinary case to be illusory. By the ordinary case I mean radiation induced cancer. Since, at low levels of radiation, the risk that any particular individual will die of a radiation induced cancer are small and since it is impossible to determine whether a cancer is radiation induced or natural, it follows that it will be virtually impossible for a person who has been subjected to radiation and has contracted a cancer to establish that the cancer was due to radiation and thus that he is entitled to compensation'.

The evidence supports the pessimistic views of Justice Parker, and those who have brought lawsuits for health damage caused by radiation have had very little success. Therefore, only a tiny proportion of the negative externalities fall upon the producers of radiation via the legal process. The polluter may voluntarily make compensation payments to the victims but, since such payments reduce the profits of the polluter, there are few such cases.

11.6 The Overproduction of Radiation

The typical allocation of property rights is that, subject to the law, producers have the right to pollute the public. In consequence the level of activity of processes that generate pollution will be too high from a social point of view because a profit maximizing producer will ignore the negative externality he or she is imposing on the public i.e. costs are understated due to market failure. This means that, without government intervention, there will be a considerable overproduction of such pollution. This is illustrated in figure 11.1, where the marginal cost curve being used by the decision maker includes only private costs. The resulting cost curve is an understatement of the true marginal social cost curve. The difference between these two marginal cost curves represents the negative externality imposed by the producer upon society. The use of understated costs leads to the level of activity of the radiation-generating activity being increased to OE and the net social benefit falling to KL. Similarly, a lack of safety measures can result in negative externalities being imposed upon those who are killed or injured as a result. Again there will be market failure, and an underuse of safety measures resulting in an overproduction of radiation.

In the UK much of the nuclear industry is state owned e.g. the Ministry of Defence design , manufacture, store and use nuclear weapons; the CEGB operate nuclear power stations; BNFL run the nuclear reprocessing facility at Windscale (Sellafield); and most medical exposure is conducted by the National Health Service. Such state organizations can be directed to consider the estimated net benefit to society of their nuclear activities i.e. to consider both the traded and non-traded effects when reaching a decision on their level of activity and on the safety measures they employ.

In other countries a much smaller proportion of the nuclear industry may be government owned e.g. in the USA nuclear power stations are privately owned whilst medical exposure is in the hands of the private sector. The aim of private sector companies is usually taken to be profit maximization. This implies considering only traded costs and benefits i.e. items that affect their profits. Such companies have no incentive to consider the non-traded costs they impose on others e.g. the cancers they cause amongst the population living near a nuclear power plant. It is both to prevent the private sector from imposing substantial non-traded costs on the population, and to give guidance to the state sector in choosing their levels of operation and safety measures, that governments have set radiation exposure limits. In order to establish an acceptable exposure standard the state must compare the social costs and benefits i.e. conduct a cost-benefit analysis.

11.7 Different Standards for Different Activities

An important error runs through much of the discussion of the social costs and benefits of radiation-generating activities. Many discussions are conducted as though there were only a single activity that generated radiation. Often nuclear power stations are used as the single radiation-generating activity. In reality, of course, there are many different activities which generate radiation. These various activities are associated with very different social costs and benefits. For example, the benefits of using X-rays for medical diagnosis and treatment are very different from those associated with the manufacture of nuclear warheads. Each radiation-generating activity must be judged separately. This implies that the socially optimal level of radiation exposure will probably differ between types of radiation-generating activity. Therefore, the radiation exposure standards also ought to differ according to the type of activity concerned. For example, Mole(1979) has stated that

> 'whether the use of radiation is for medical or industrial exposures it is usual to set some dose level which should not be exceeded. This will be zero for an industrial use with a benefit which is regarded as trivial or when the benefit can be obtained by other means with a smaller concomitant cost (including risk)'.

This result, that exposure standards should differ between types of activity, has been largely ignored by those bodies that set radiation exposure standards. The exposure standard is the same for all workers regardless of whether they are manufacturing warheads or operating a hospital X-ray machine. Similarly, the exposure standard for members of the public is the same for leaks from nuclear submarines and for radiation from TV sets. The only area in which a distinction has been made between different types of exposure is for patients undergoing medical procedures. In this case there is no upper limit on the exposure, and it is up to the doctor to balance costs against benefits. Proper cost-benefit analysis implies that there should be a range of exposure standards, with a separate standard for each type of radiation-generating activity. This will require the costs and benefits of each type of activity that produces radiation to be investigated in detail. It would mean that radiation exposure caused by the manufacture of nuclear weapons would have to be justified in terms of the resulting benefits, and it could no longer hide behind other, more beneficial, sources of radiation exposure.

11.8 Measuring the Social Costs and Benefits

Before discussing the measurement of social costs and benefits, it is worthwhile noting that whether some non-traded effects are classified as a cost or a benefit is a matter of opinion. For example, some people may regard the explosive properties of a nuclear warhead as a benefit whilst others might regard it as a cost. Of course, once a non-traded effect has been valued it can then be classified as a social cost or benefit according to whether its overall value to society is positive or negative. However, the classification of some effect as a cost or benefit is unimportant. Thus, whether the effects of safety measures are categorized as an increase in social benefits or a reduction in social costs (as was done earlier in this chapter) does not affect the optimal level of safety measures. In consequence, it will simplify matters if the social costs and benefits of some radiation-generating activity are lumped together and just described as the effects.

If the effects on society from some activity are to be compared, three steps are necessary. First, the effects of the radiation-generating activity upon society must be identified. Some of these effects are not immediately obvious. For example, the production of electricity by nuclear power stations involves cancer amongst uranium miners; health risks to children playing on the tailings; radon exposure for those living in homes made from uranium tailings; deaths for workers in nuclear fuel fabrication, reprocessing and waste treatment plants; and the pollution of the environment with radioactive waste. There is also the problem that some linkages are concealed.

The authorities try to present nuclear power generation and nuclear weapons as being separate activities whilst, in fact, they are inextricably linked. This interaction between the military and civil uses of nuclear technology makes the identification of the effects of any particular radiation-generating activity more difficult. At the Sizewell inquiry, in January 1983, the CEGB claimed that no plutonium produced by the CEGB reactors had been used for the production of nuclear weapons either in the UK or elsewhere. But shortly afterwards Lord Hinton, the first chairman of the CEGB, described this claim by the CEGB as 'bloody lies', quoted from Hesketh(1984). Lord Hinton's comment implies that plutonium from civil reactors has been used to make nuclear weapons, and that there is a direct link between the civil and military nuclear programmes in the UK.

There is also a linkage between the civil and military nuclear programmes at Windscale (Sellafield). Over the period 1974-1984 about 15% of the liquid discharges were caused by military uses of the plant, Black(1984, page 43). One effect of this military use of Windscale (Sellafield) is that nuclear safety inspectors are barred from parts of the plant because inspection by them may lead to the revelation of classified information on the production of military material. For similar reasons the Holliday inquiry was unable to discover how much of the radioactive waste dumped in the North Atlantic by the UK comes from the weapons programme, Holliday(1984, page 53-54). Brindle et al(1983) have estimated that about one third of the waste prepared for sea dumping by the UK in 1983 came from the MoD. Since some of the linkages between the military and civil nuclear programmes are secret, it is doubly difficult to identify the effects of some radiation-generating activity.

Second, the effects that have been identified must be quantified e.g. number of people killed, tons of radioactive waste dumped in the sea, lives saved by accurate diagnosis, number of wars deterred by nuclear weapons, quantity of electrical power produced, tons of uranium used, etc. Third, these social effects must be converted to some common unit in order that the net effect upon society can be calculated. For example, a programme of mass X-ray screening for breast cancer will save some lives due to early diagnosis whilst the X-rays of the breast will induce some cancer deaths. So 'lives' could be used as the common unit of measurement, and other costs and benefits converted into lives. However, in almost all cases, money is chosen as the common unit in which to express the social costs and benefits. This is, in large part, because all the private costs and benefits generated by the activity are already in monetary terms e.g. the units of electricity sold, salaries paid to employees etc. In consequence. the problem resolves into placing monetary valuations on the non-traded costs and benefits.

An example of the estimation of some non-traded costs associated with the Three Mile Island accident is provided by Hu et al (1984). This accident, which is discussed in chapter 9, resulted in negative externalities for the local residents, 150,000 of whom were evacuated. Hu et al investigated the extra expenditure caused by stress related to the accident. They studied the effects upon those living within a five mile radius of the plant for ten months after the accident. Survey data was used to estimate the number of extra visits to the doctor and work days lost, and also the increase in the consumption of sleeping pills, tranquilizers, alcohol and cigarettes. The average market prices of these items were then used to calculate the monetary cost to the local residents of the stress caused by the accident. This resulted in an estimated cost of $180,000 over a ten month period. This figure ignores many other costs imposed upon the local population such as miscarriages, stillbirths and cancers induced by the accident. In this example the difficult step was in quantifying the effects of the stress upon the economic behaviour of local residents. Converting these various changes of behaviour into a common unit, money, was straightforward because market prices were readily available. Those effects which would have been more difficult both to quantify and to value were excluded from the study. This illustrates a common danger with cost-benefit studies, which is that only those costs and benefits which are easy to express in monetary terms are included.

Quantifying the non-traded effects on society of some radiation-generating activity is often a hard task. Consider the health effects of radiation. The investigator must first quantify the radiation dose received by humans. For various medical procedures there may be reasonable data on which to base such an estimate, but where radioactive material is released into the environment the problem of estimating the dose to humans is extremely difficult, see chapter 10. Even supposing an accurate estimate can be made of the radiation dose, there is still the controversial issue of the dangers of low level radiation. It follows that, if the estimated dose is multiplied by an estimated risk factor, the resulting figure for the number of adverse health effects may be highly inaccurate. This problem was encountered when considering the dangers of Windscale (Sellafield) to the health of the local population.

Whilst some effects are potentially quantifiable with an obvious choice of units e.g. cancer deaths, others do not have any obvious units in which they can be measured, and therefore are much more difficult to quantify.

Thus, the effects of mounds of uranium tailings or the cooling towers of nuclear power stations may affect the scenic quantity of the area, an expansion of nuclear power may help in the defeat of a miners' strike, nuclear weapons may provide deterrence and also the means for ending the existence of the human race, the protection of the nuclear industry may lead to the government deliberately deceiving the public etc.

Once the investigator has identified and quantified the effects of some radiation-generating activity, he or she must then place a monetary valuation on these different effects so that the costs and benefits can be summed. In some cases valuation may be a speculative process e.g. placing a monetary value on deterring a nuclear war or on the value of a human life. The thorny problem of valuing a human life will be considered in detail in the next section. As if these problems were not enough, many of the effects of radiation-generating activities last for a large number of years e.g. plutonium released into the environment will emit deadly radiation for tens of thousands of years. This raises questions such as whether a cancer death now has the same social cost as a cancer death in ten thousand years time. This problem of the effects being spread out over many years can be solved by the use of a social discount rate. However, there is no agreement on the appropriate figure to use for the social discount rate, and so a further area of controversy is added to the many that have already been discussed.

11.9 The Value of Human Life

Some people may take the view that human life is priceless and so cannot have a monetary value placed upon it. But, whether we like it or not, society makes decisions which implicitly do just this. For example, every time the UK government decides not to spend money on a kidney machine that would keep alive a person with kidney failure, they are effectively saying that the life of the patient is worth less than the cost of providing the machine. In the context of radiation, the government has decided that the benefits from the manufacture of nuclear warheads at the AWRE Aldermaston exceed the costs, including the likely cancer deaths downwind of the plant caused by the release of plutonium dust.

Valuations of human life have been expressed in two different ways, (a) value per man-rem, and (b) value per death. An estimate of the probability of death if one person is exposed to one rem (ten mSv) is required to convert a value per man-rem into a value per death. Thus, the value per death is equal to the value per man-rem divided by the risk of death if one person is exposed to one rem (ten mSv). Various valuations of the exposure of the public to a man-rem of radiation are shown in table 11.1.

Table 11.1 Valuations of Human Life

Organization	Valuation per Man-rem	Valuation per Life	Reference
NRC	$1,000	$1,670,000	Roberts(1984, p79)
MAFF	£50	£80,000	Taylor(1982, p19)
NRPB -			Clark et al(1981c,p6)
0.000 to 0.005 rem	£20	£30,000	"
0.005 to 0.050 rem	£100	£170,000	"
0.050 to 0.500 rem	£500	£830,000	"
EPA	£500	£830,000	Taylor(1982, p19)

This shows that there is considerable variation in the value given to human life, even between official organizations e.g. two orders of magnitude. Even if allowance is made for the fact that these valuations were made at different times in the last decade, substantial differences still remain.

Whilst the valuations in table 11.1 were expressed in terms of a value per man-rem, they can easily be converted to a value per life. This has been done in the right-hand column of table 11.1 using a risk factor of six hundred deaths per million people exposed to one rem (ten mSv). Clearly other risk factors would produce different implied values for a human life, see table 5.2 in chapter 5 for some other possible risk factors. The greater the risk the smaller is the implied value of human life. The values in table 11.1 are for the average human life, and the value placed on a particular life may vary according to life expectancy etc.

The valuation of human life proposed by the NRPB depends on the size of the radiation dose received. This means that, in order to place a value on the risk of death caused by some radiation dose, it is necessary to know the distribution of this dose amongst the people concerned. In some cases it may be difficult to estimate the distribution of the dose. In such cases the NRPB have proposed that the average dose be used, Clark et al (1981c, page 36). The numbers presented in table 11.1 imply that the NRPB's valuation function is concave. Therefore, for the reasons given in figure 4.3 of chapter 4, the valuation of the average dose will be less than the average of the valuation of the individual doses i.e. the use of average doses will tend to bias downwards the resulting valuation of the risk of damaging human life.

Curiously, Clark et al (1981a) of the NRPB have stated that their assumptions imply that the 'individual valuations of risk changes will rise at an increasing rate' and this means that the valuation function will be convex. Indeed, Clark et al (1981b) contains a graph of a valuation function that is convex. Similarly, the NRPB(1980, page 69) refers to the 'increasing marginal valuation of risk'. But, as mentioned above, the NRPB's actual valuation function for valuing human life is concave! The use of a concave valuation function implies that individual valuations of risk rise at a decreasing rate and, if extrapolated, that there is some sum of money at which people are prepared to accept certain death! Such a proposition is open to considerable criticism.

Since the NRPB's valuation of radiation risk varies according to the dose, an important question is how the relevant dose is calculated. The NRPB view is that, not only should background radiation be excluded from such a calculation, but that *all* radiation, other than that from the particular source being analyzed, should be disregarded, Clark et al (1981c, page 38). This NRPB view, which was adopted for purely practical reasons, raises a number of problems. First, there is the issue of defining what is an independent source. Is Sizewell B nuclear power station independent of Sizewell A for these purposes? Second, the NRPB's approach assumes, without any justification, that people value risk from different radiation-generating activities independently. A much more plausible approach is to regard a person's attitude towards risk as a function of their entire radiation dose. For example, a member of the public may be exposed to 0.003 rem (0.03 mSv) per year from each of two independent sources of radiation, making a total dose of 0.006 rem (0.06 mSv) per year. This suggests that a valuation of £100 per man-rem be applied. However, the NRPB would only value this dose at £20 per man-rem.

Two main alternative approaches to placing a value on human life have been used in connection with radiation exposure. These are (a) calculating the real resource effects of a death e.g. medical costs, loss of net output etc., and (b) the total of the amounts of money required by each person to compensate them for an increased risk of death. Another method for valuing human life involves the use of data on wage rates and market prices to calculate the implicit valuation of a human life. This approach can produce estimates that vary from $400,000 to $24.2 million in 1982 prices, i.e. a difference of sixty times, Smith et al (1984). This market based approach is very sensitive to the assumptions made about the dose-response curve, and has not been applied to radiation exposure.

Jones-Lee(1981) argues that the correct way to value a human life is the sum of each individual's valuation of a change in the risk of their own death, i.e. (b), plus the real resource effects not already included in (b). The NRPB(1980, page 5) also take this view. Jones-Lee then argues that individual valuations of a change in risk are much larger than the real resource costs and he goes on to ignore the real resource costs in his analysis.

He also points out that, in order to place a valuation on exposure to some radiation dose, each person is required to make his or her own estimate of the resulting risk of death i.e. subjective probabilities. Thus each person must make his or her own estimate of the dose-response curve! The NRPB have exploited this use of subjective probabilities to argue that, although there is a positive risk of death from a very small dose, people's subjective estimate of the risk of death is zero. This is illustrated in figure 11.3 where, if the true risk is less than OA, the corresponding subjective risk is zero. Above OA it has been assumed that the subjective and true risk estimates are equal, and so the curve is OABC. This means that the value to be placed on exposure to very low levels of radiation is just the real resource costs, NRPB(1980, page 10). If a population of n people is exposed to one man-rem (0.01 man-sieverts) the average dose is 1/n rem (10/n mSv). As n becomes very large the average dose per person will become small enough for it to be argued that the subjective valuations are zero. Professor Jones-Lee has criticised this NRPB approach and stated that 'a methodology which implies that the value of statistical life can be driven close to zero provided that the population is large enough is, to say the least, suspect', Jones-Lee(1981). Jones-Lee's solution to this difficulty is to use risk estimates supplied by experts i.e. objective probabilities.

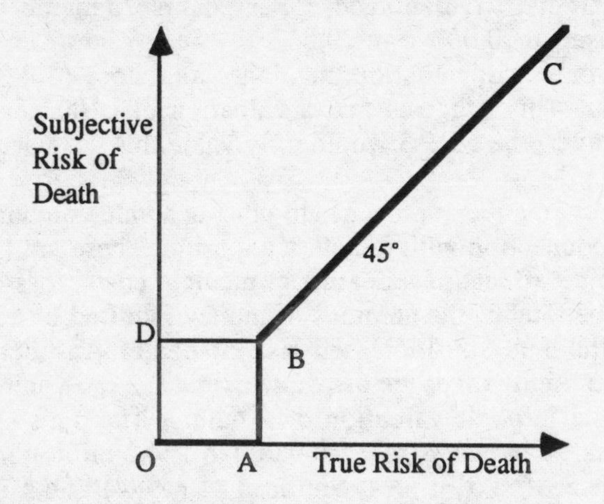

Figure 11.3 Relationship Between True and Subjective Risk

An alternative approach to this difficulty is based upon the argument, often put forward by the nuclear industry, that the public have an irrational fear of small amounts of radiation and grossly *over*estimate the risks. For example, it has been claimed that the major health effects of the Three Mile Island accident were psychological rather than physical. If people do grossly overestimate the risks from small amounts of radiation then, rather than assigning zero risk to low doses, they may assign some minimum risk to all additional radiation exposure e.g. OD in figure 11.3. In this case the curve becomes ODBC and the NRPB's case for largely ignoring low doses per person disappears.

An alternative argument for ignoring small doses is that there is some minimal dose below which the risks cannot be detected, e.g. Adler et al (1978) and Comar(1979). Therefore, doses below this level can be ignored. This is a version of the 'threshold dose' argument, which was disposed of in chapter 2.

There are advantages to both the regulatory authorities and the nuclear industry if small doses are ignored. Such an approach simplifies the calculation of the collective dose by restricting both the number of people and the time period concerned. In accordance with ICRP advice, radiation doses to UK workers of the order of 0.1 rem (1 mSv) per year are not measured and rarely assessed, Fleishman (1985, page 4). Thus data on these doses need not be collected by the employers nor inspected by the regulatory authorities. For members of the public Fleishman (1985, page 7) has proposed that, if individual and collective doses are less than some minimal amount, the exposure will automatically be regarded as insignificant.

Some previous estimates of the value of human life have been based on health expenditure i.e. one component of the real resource effect. For example, Cohen(1973) implicitly calculated the average value of ill-health using total US consumer and government expenditure on medical care together with an estimate of the total number of Americans subject to ill-health. This is a most unsuitable way of placing a value on human life (or ill-health). There need be only a slight connection between the average amount spent by a private enterprise health system on treating ill-health and the value placed by society on a human life (or ill-health). For example, expenditure on cosmetic surgery, dentistry and spectacles is included in Cohen's calculation whilst very little may be spent treating someone who is killed in a car accident or who is stillborn.

It is interesting to compare the valuations of human life in table 11.1 with the compensation and damages actually paid by employers in respect of employees who have been killed by cancer. In chapter 8 it was reported that BNFL have paid £21,654, £31,000 and £120,000 to the dependents of three workers. In the US the family of Karen Silkwood was awarded $10.5 million damages, Rashke(1982, page 378), while nine families have been awarded a total of $2.66 million for the adverse health effects of the Nevada bomb tests.

It is sometimes argued that the protection given to workers and the general public from radiation exposure is excessive, in that it is higher than for any other health risk. However, at least in the US, this is clearly not the case. The US legislation on the carcinogenic effects of food additives is very strict, and the Delany Amendment of 1958 bans a food additive if it has *any* carcinogenic effects. Thus, the same people who are protected absolutely so far as food additives are concerned, have only a partial protection from radioactive emissions. In other words it is acceptable in the US if some people die from cancer caused by radioactive discharges, but no-one must be allowed to die from cancer caused by food additives.

It may be argued that the higher radiation risk for nuclear workers is a traded rather than a non-traded effect. Thus, in a competitive labour market the wage rate of nuclear workers is argued to include a sum to compensate for the additional radiation risk. But, this presumes both a competitive labour market and that would-be nuclear workers are aware of the health risks involved. Neither of these conditions is met. Labour markets are not competitive in the sense required here, and workers are typically ill informed about the health risks they run. Gofman(1983, page 355) has written that 'workers are encouraged directly and indirectly to believe that permissible means safe', i.e. that nuclear workers are not subject to any additional risk. The evidence presented in chapter 8 makes it clear that nuclear workers do run additional health risks from their exposure to radiation, and that these health risks may well be larger than is officially admitted. For these reasons, it is argued that the major part of the additional health risk from becoming a nuclear worker is not traded.

Earlier in this chapter, it was argued that making socially optimal decisions about the level of activity and safety measures for radiation-generating activities necessitates a cost-benefit approach. Cost-benefit analysis obviously requires that both the costs and the benefits be considered. However, major bodies have managed to set exposure standards for both workers and the public without *any* direct consideration of the benefits. What has been done is to use the health risks associated with other occupations or public activities. It is argued that, because these other risks are permitted, so should a similar risk due to radiation exposure. Such a procedure does not constitute cost-benefit analysis because the benefits of the radiation-generating activity have not been considered. For example, because thousands of people are killed in road accidents each year does not mean that *any* activity which kills a similar number of people is justified.

In setting public exposure standards the ICRP(1977, page 23) stated that

> 'the acceptable level of risk for stochastic phenomena for members of the general public may be inferred from considerations of risks that an individual can modify to only a small degree and which, like radiation safety, may be regulated by national ordinance. An example of such risks is that of using public transport.'

Thus, the benefits from activities such as public transport are effectively being substituted for the benefits from the radiation-generating activity in the cost-benefit calculation. Similarly, when considering occupational exposure standards, the ICRP(1977, page 19) stated that 'the Commission believes that for the foreseeable future a valid method for judging the acceptability of the level of risk in radiation work is by comparing this risk with that for other occupations recognized as having high standards of safety.' So, it is the benefits from other types of occupational exposure that are being used to set exposure standards for radiation workers.

When considering public exposure from Windscale (Sellafield), Black(1984, page 88) states

> 'that the question to be considered is whether the discharges from the Sellafield site pose a greater hazard than other imposed, hidden, low exposures to long term hazards normally accepted by the public. In order to consider this question, it is necessary to find out what risks are generally found to be acceptable by members of the public.'

Black(1984, page 89) then asserts that 'extensive monitoring has indicated that the increased radioactivity (from Windscale (Sellafield)) in the general environment is of an order which is accepted in other situations.' Black has not considered the benefits received by society from the operation of Windscale (Sellafield), yet manages to imply that the emissions are justified. What Black has done is to attribute the benefits associated with other exposures to the radioactive emissions from Windscale (Sellafield).

Reissland et al (1979) of the NRPB have written that 'the benefits of nuclear power are self-evident' and 'we will not attempt to quantify the benefits.' They then go on to consider the health risks from a variety of occupations, and point out that the exposure standards for nuclear workers result in lower risks than in a number of other industries. Again, the implication is that if the occupational risks in these other industries are justified, then so are the risks for nuclear power workers. However, just because deep sea fishing is a very risky occupation does not mean that the risks of being a nuclear power worker are justified.

On the basis of the level of fatal risks that are acceptable to the general public in everyday life, the ICRP(1977, page 23) conclude that a public exposure standard of 0.1 rem (1 mSv) per year is implied. However, the ICRP has decided to permit the general public to be exposed to five times this level of radiation i.e. 0.5 rem (5 mSv) per year. Their reasons for setting an exposure standard that is 500% larger than implied by their own calculations appear to boil down to :

(a) in reality very few members of the public will actually be exposed to 0.5 rem (5 mSv) per year,

(b) exposure to 0.5 rem (5 mSv) per year over a long period of time would not lead to risks appreciably above the level that is generally considered acceptable (although a linear dose-response curve implies a five fold increase),

(c) the ALARA concept will tend to restrict doses anyway, and

(d) the previous public exposure standard was 0.5 rem (5 mSv) per year.

The ICRP's rationale for a five fold increase in the public exposure standard indicated by their data is most unconvincing. If, in practice, no member of the public will be exposed to 0.5 rem (5 mSv) per year there can be no objection to setting a tougher standard. The ICRP appear to be seeking a way of justifying the existing exposure standard, possibly because a tougher standard might create problems for present nuclear plants. In doing this the ICRP have to abandon their own inadequate version of cost-benefit analysis.

11.10 Distribution of the Social Costs and Benefits

An important aspect of the non-traded costs and benefits of some radiation-generating activity is their distribution amongst people. For example, the cancers induced by radioactive emissions from Windscale (Sellafield) and Aldermaston fall most heavily on those living near the plant. To the extent that there are benefits to society from these plants, they are received by the nation and not just by the local population. Therefore, even if the operation of the plant results in a net benefit to society, some people may be made worse off e.g. those who die of cancer. This change in the distribution of welfare is yet another non-traded effect that must be incorporated in a cost-benefit analysis. Of course, quantifying and valuing a change in the distribution of welfare is a very difficult task.

The distribution of the non-traded costs of a radiation-generating activity (e.g. radiation health risks) may be unequal when analyzed by a variety of factors. As already shown, those people living close to nuclear plants or nuclear weapons testing areas are subject to higher risks. It was stated in chapter 1 that a child is more susceptible than an adult to radiation health risks, with the foetus being the most sensitive of all. In addition to being more likely to contract cancer, young people have a greater life expectancy and so lose more years of life when killed by cancer. For these reasons, the young bear a higher share of the costs than the old. In some cases the introduction of safety measures will alter the distribution of the radiation dose amongst the population. Thus, the additional treatment of radioactive waste will increase the already above average radiation dose received by nuclear workers, and reduce the dose to the public, Clark et al (1981c). An implication of differences in the health risks between different groups in society e.g. the young and the old or males and females, is that, assuming the social benefits are substantially the same, the exposure standards ought to be different for each of these groups. But, as Matthews(1983) has pointed out, differential exposure standards may lead to job discrimination, e.g. women of child bearing age being rejected as radiation workers in favour of older women.

For medical exposure to radiation, most of the costs and benefits are received by the patient, and so the distributional effects are small.

Conversely, as described in chapter 7, the population of some of the Marshall Islands has suffered very considerably as a result of US nuclear weapons tests, whilst such benefits as there may be from testing nuclear weapons have been received by the entire US population. In this case there has been a clear distributional effect.

A possible way of dealing with distributional effects is to try to compensate those who would otherwise be made worse off. The result is that there are no losers, and the change in the distribution of welfare is much reduced. In a few cases the polluter may voluntarily pay compensation to the victims of their pollution. Thus, in Japan substantial compensation payments are made to people living close to nuclear power stations. In addition, these plants have to pay higher local property taxes for the first fifteen years of their life. This is also of benefit to the local community since it reduces the revenue that must be raised from local residents, Fishlock(1984b). In France people living near nuclear installations receive a reduction in electricity prices, Fishlock(1984a).

Whilst the payment of compensation to those who suffer from non-traded costs tends to reduce the change in the distribution of welfare, it is not possible to ensure that no-one is made worse off. There are a number of reasons for this difficulty. Suppose the government asks each person to state how much money they require to compensate for the non-traded costs imposed on them. People will have a clear incentive to overstate grossly the monetary sum required to compensate them, and the resulting numbers may well be so large that the government will refuse to make the payments. Another problem is that compensation schemes typically set a common payment for all those people in the same circumstances. But, even if everyone answers the government's question honestly, people in the same circumstances will probably give different answers. This means that it is impossible to exactly compensate everyone, and to ensure that no-one loses out, the common level of compensation will have to be set very high.

Since compensation payments are unlikely to take place, radiation exposure standards have been set to place an upper limit on the dose to which any person should be subjected. In some cases these limitations on individual exposure just mean that the same total dose is distributed among more people e.g. the practice of 'burning out' workers described in chapter 8. In other cases, where the dose to any particular individual cannot be controlled, emissions must be limited so as to prevent the dose to the critical group exceeding the exposure standard.

11.11 Summary

All exposures to man-made radiation are preventable and the extent to which people are exposed to such radiation is a matter of choice. This chapter has outlined a cost-benefit approach to making three interrelated decisions, (a) what production technology should be used in the radiation-generating activity, (b) what is the socially optimal level of operation of the radiation-generating activity, and (c) what is the socially optimal quantity of safety measures? A competitive price system will fail to make socially optimal decisions on these three matters. Profit maximizing companies will ignore the costs that they are imposing on others e.g. by polluting the environment with radioactive material. In consequence, there will be an overproduction of radiation relative to the socially optimal level. In an attempt to correct for this inbuilt incentive to oversupply radiation, the authorities have set exposure standards.

Since the social benefits clearly differ between radiation-generating activities, so should the exposure standards. However, this is not the case. The exposure standard for a radiologist who is engaged in diagnosing or treating illness is the same as for a worker at the AWRE Aldermaston manufacturing nuclear warheads. Although the authorities now appear to accept that a cost-benefit analysis is appropriate to setting radiation exposure standards, some of their current techniques give cause for concern. For example, the way in which valuations have been placed upon human life. Since a cost-benefit study involves many items that are almost impossible to value in a generally acceptable manner, and since radiation exposure standards depend upon such valuations, it follows that almost *any* radiation standard is open to challenge. Cost-benefit analysis will not produce definitive answers; all it does is to provide a helpful framework for the debate over standards.

Glossary of terms

Absorbed Dose. This is a quantity of ionizing radiation and is measured as the amount of energy imparted to a unit mass of matter such as tissue. It used to be measured in rads and is now measured in grays.

Actinides. The group of fifteen elements with atomic numbers 89 to 103, which includes uranium and plutonium.

Air Dose. The dose that the tissue would receive if it were surrounded by air and not shielded by intervening body tissue.

Alpha Particle. A particle consisting of two protons and two neutrons which has a double positive charge and a mass that is four times that of a proton.

Americium (Am). An artificially created radioactive element.

Anencephalus. A congenital abnormality in which the brain is absent.

Ankylosing Spondylitis. Inflammation of the vertebrae of the spine leading to the complete fixation of the spine as a result of the bones fusing together.

Background Radiation. Naturally occurring cosmic and geological radiation. Sometimes radiation from nuclear weapons fallout may also be included.

Becquerel (Bq). See curie.

Beta Particle. A particle with a mass and charge equal to those of an electron.

Burning Out. See glowboy.

Cancer. A malignant tumour that invades and destroys the surrounding tissue.

Carcinogen. A substance that causes cancer.

Chemotherapy. The treatment of a disease using chemicals.

Chromosome. A rod-shaped body found in the nucleus of every cell in the body. It contains the genes, or hereditary information. Humans have 46 chromosomes per cell.

Collective Effective Dose Equivalent. This is calculated by multiplying the average effective dose equivalent by the number of people exposed. It is expressed in units called man-rems or man-sieverts and is often abbreviated to collective dose.

Concave Function. Viewed from below the curve appears concave.

Cosmic Ray. Ionizing radiation from outer space which is a mixture of various types of radiation such as gamma rays.

Cost-Benefit Analysis. An economic technique for decision making which involves a comparison of the costs and benefits to society.

Curie (Ci). A measure of the amount of radiation given off by a radioactive substance. The radioactivity of one gramme of radium is a curie. It has recently been replaced by the becquerel, where one becquerel is equivalent to one radioactive disintergration per second.

DNA. The compound that controls the structure and function of cells, and is the material of inheritance.

Dose Equivalent. The absorbed dose is multiplied by a factor to allow for differences in biological effectiveness between different types of radiation. It used to be measured in rems and is now measured in sieverts.

Dose Fractionation. A given total dose of radiation is administered in relatively small doses daily or at longer intervals.

Dose Protraction. A dose of radiation is delivered continuously over a relatively long period at a low dose rate.

Doubling Dose. The amount of radiation required to double the natural incidence of a health effect.

Disintegration. The emission of alpha, beta or gamma rays.

Effective Dose Equivalent. This is obtained by multiplying the dose equivalents to the various parts of the subject's body by the appropriate risk weighting factors and summing the results. It is expressed in sieverts and is often abbreviated to dose.

Electron. A particle with a small mass (1/1836 that of a proton) that has a negative charge.

Empirical Study. A study which is founded on observations (either of the real world or of experiments) and not on theory.

Epicentre. See ground zero.

Epidemiology. The study of disease occurrence in human populations.

Excitation. The impartation of energy to an atom which is dissipated as heat or light without causing ionization.

Fluoroscopy. The subject is continuously exposed to X-rays which are projected onto a screen. This enables movements within the body to be observed.

Free Radical. A grouping of atoms that normally exists in combination with other atoms, but which can sometimes exist independently, and is generally very reactive in a chemical sense.

Gamma Ray. A quantity of energy, without mass or charge, that is propagated as a wave.

Genetic Effect. An effect which appears in the descendents of the exposed subjects.

Germ Cell. The parent cell from which a new individual develops.

Gray (Gy). See absorbed dose.

Glowboy. An unskilled, short-term employee who exposes himself to relatively high doses of radiation for a few minutes. In this time they are 'burned out' by receiving the maximum permissible dose for a much longer period of time.

Granulocytic Leukaemia. See leukaemia.

Ground Zero. That point on the Earth's surface directly below (or above) the center of the explosion.

Gut Uptake Factor. The proportion of a substance in the gut that is absorbed into the body.

Half-life. The time taken for half of the initial quantity (by weight) to decay to its daughters, which may also be radioactive.

Health Physics. The function of studying and advising on all matters relating to radiation safety and ensuring that safety standards are observed.

Hot Particle. An alpha-emitting particle in the deep respiratory tissue of such activity as to expose the surrounding lung tissue to a dose of at least 1,000 rem (10,000 mSv) per year.

Hot Spot. See hot particle.

Intravenous Pyelogram. An X-ray examination of the kidney and the upper urinary tract.

In Vitro. Studies carried out on material from an animal under artificially controlled conditions in the laboratory. Literally 'in glass'.

In Vivo. Studies carried out on a living animal. Literally 'in life'.

Ionization. The process by which a neutral atom or molecule acquires an electric charge. An atom acquires a positive charge when it loses electrons and a negative charge when it gains electrons.

Irradiate. To subject to radiation.

Isotope. All naturally occurring elements have radioactive isotopes. Isotopes are chemically the same as the corresponding element, but have a different number of neutrons and therefore a different mass.

Jumper. See glowboy.

Kerma. Kinetic Energy Released in Material: this is the value of the T65D dose.

Latent Period. The time period which elapses between the event which causes the cancer and the diagnosis of the disease.

Leukaemia. A rare disease where large numbers of immature and abnormal white blood cells appear in the blood. It can be classified by the dominant type of cell involved into granulocytic, lymphocytic and monocytic.

Linear Energy Transfer (LET). The average amount of energy lost per unit of distance travelled.

Low Level Radiation. A small cumulative dose of radiation e.g. under ten rem.

Lymphocytic Leukaemia. See leukaemia.

Mammography. X-ray examination of the breast.

Man-rem. See collective effective dose equivalent.

Marginal Cost or Benefit. The amount by which the total cost or benefit rises when the level of activity is expanded by one unit.

Myeloma. A rare disease involving the proliferation of malignant plasma cells within the bone marrow.

Myeloid Leukaemia. See leukaemia.

Neoplasm. Any new and abnormal growth such as a cancer.

Neutron. An elementary particle with approximately unit atomic mass and no electric charge.

Non-stochastic Effect. An effect which is certain to occur, with the size of the effect varying in both frequency and severity as a function of the dose.

Nucleus of a Cell. The central part of a cell which contains the crucial DNA.

Nuclide. Any isotope of an element.

Parameter. A constant quantity entering into the equation of a curve.

Phospholipid. A fat containing phosphoric acid. They are especially common in the brain and nerve sheaths.

Pitchblende. An oxide of uranium that is a source of uranium and radium.

Plutonium (Pu). A radioactive element that is artificial, apart from a small quantity in West Africa created by a natural nuclear reaction.

Private Costs. Costs which involve actual monetary payments.

Proton. An elementary particle with approximately unit atomic mass and unit positive electrical charge.

Rad. See absorbed dose.

Radiobiology. The study of the effects of radiation on living things.

Radiography. A single instantaneous X-ray exposure is recorded on a photographic plate.

Radiological Protection. The science and practice of limiting the harm to humans from radiation.

Radionuclide. A nuclide that is in an unstable form so that it emits ionizing radiation.

Radiotherapy. The use of radiation to kill off malignant cells.

Radon. An inert radioactive gas produced naturally by the decay of radium, thorium and other actinides.

Recombination. A collision between free radicals which converts them into uncharged atoms.

Relative Biological Effectiveness (RBE). For a particular type of radiation it is equal to the number of rems divided by the number of rads. For X-rays, gamma rays and beta particles the RBE is equal to one.

Rem. Roentgen equivalent man, see dose equivalent.

Reprocessing. The extraction of uranium and plutonium from the spent fuel of a nuclear reactor.

Reticuloendothelial System (RES). A system of highly specialized cells scattered throughout the body, but found mainly in the spleen, bone marrow, liver and lymph glands. Their main function is the ingestion of red blood cells and the conversion of haemoglobin to bilimbin. These cells also ingest bacteria and foreign colloidal particles.

Roentgen. A unit of radiation exposure that was replaced by the rad.

Sarcoma. A malignant tumour composed of connective tissue in the bones, muscles, sinews etc. They are very much less common than cancers.

Scintillator. A chemical that glows when exposed to radiation.

Sellafield. The BNFL site in West Cumbria which includes the Windscale nuclear fuel reprocessing facility and the Calder Hall nuclear reactors.

Sievert (Sv). See dose equivalent.

SMR. The rate of death for a particular area after standardizing for age, expressed as a percentage of the overall national rate.

Stochastic Effect. An effect which may or may not appear in any particular individual i.e. probabilistic. It varies in frequency but not severity as a function of the dose, without any threshold e.g. cancer.

Social Cost. The total cost to society i.e. the private costs plus the negative externalities.

Somatic Effect. An effect that appears in the exposed subject as opposed to genetic effects which appear in the descendents of the exposed subject.

Supralinear. See concave function.

Synergy. The combination of factors produces a result which is greater than the sum of the parts.

Teratogenic Effect. The production of physical defects in the foetus e.g. the malformed limbs produced by thalidomide.

Thermoluminescent. A material that emits light when it is heated.

Thorium (Th). A radioactive element that decays to thoron, an inert radioactive gas.

Thorotrast. A substance containing thorium which was injected into patients as an X-ray contrast medium, primarily for the diagnosis of suspected brain diseases.

Threshold Dose. The minimal absorbed dose that will produce a detectable degree of any given effect.

Transuranic. An element with atoms heavier than those of uranium.

Uranium (U). The heaviest naturally occurring radioactive element.

Uranium Tailings. An accumulation of rock and ore residues from which milling operations have extracted almost all of the uranium.

Whole Body Dose. See effective dose equivalent.

X-ray. A quantity of energy, without mass or charge, that is propagated as a wave and can be artificially produced by a machine.

Abbreviations

ABC = Associated Broadcasting Corporation
ABCC = Atom Bomb Casualty Commission
AEC = Atomic Energy Commission
ALAPA = As Low As Publically Acceptable
ALARA = As Low As ReasonablyAchievable
ALARP = As Low As Reasonably Practical
ALATA = As Low As Technically Achievable
ALL = Acute Lymphocytic Leukaemia
ANZAP = As Near Zero As Possible
AUEW = Associated Union of Engineering Workers
AWRE = Atomic Weapons Research Establishment
BBC = British Broadcasting Corporation
BEIR = Biological Effects of Ionising Radiation
BNFL = British Nuclear Fuels Ltd.
CEGB = Central Electricity Generating Board
CMD = Comparative Mean Dose
DGSE = General Directorate for External Security
DHSS = Department of Health and Social Security
DNA = Deoxyribonucleic Acid

DREF = Dose Rate Effectiveness Factor
EPA = Environmental Protection Agency
ERDA = Energy Research and Development Administration
EURATOM = European Atomic Energy Community
FAO = Food and Agriculture Organisation
GAO = General Accounting Office
GMBATU = General, Municipal, Boilermakers and Allied Trade Union
HEW = Health, Education and Welfare
HGF = Horizontal Geomagnetic Flux
HIP = Health Insurance Plan
HSE = Health and Safety Executive
IAEA = International Atomic Energy Agency
ICE = Internal Comparisons by Exposure
ICRP = International Commission on Radiological Protection
ILO = International Labour Organisation
IQ = Intelligence Quotient
ITV = Independent Television
KMS = Kneale, Mancuso and Stewart
KSM = Kneale, Stewart and Mancuso
LDC = London Dumping Convention
LET = Linear Energy Transfer
MAFF = Ministry of Agriculture, Fisheries and Food
MoD = Ministry of Defence
MP = Member of Parliament
MPBB = Maximum Permissable Body Burden
MRC = Medical Research Council
MSK = Mancuso, Stewart and Kneale
NAS = National Academy of Sciences
NBC = National Broadcasting Corporation
NCRP = National Council on Radiological Protection
NEA = Nuclear Energy Agency
NIOSH = National Institute for Occupational Safety and Health

NIREX	= Nuclear Industry Radioactive Waste Executive
NRC	= Nuclear Regulatory Commission
NRPB	= National Radiological Protection Board
NUS	= National Union of Seamen
OECD	= Organisation for Economic Cooperation and Development
OPCS	= Office of Population Censuses and Surveys
ORNL	= Oak Ridge National Laboratory
PERG	= Political Ecology Research Group
RAF	= Royal Air Force
RBE	= Relative Biological Effectiveness
RCEP	= Royal Commission on Environmental Pollution
RERF	= Radiation Effects Research Foundation
RES	= Reticuloendothelial System
RWMAC	= Radioactive Waste Management Advisory Committee
SAT	= Scholastic Aptitude Test
SKM	= Stewart, Kneale and Mancuso
SMR	= Standardised Mortality Ratio
T65D	= Temporary 1965 Dosimetry
TGWU	= Transport and General Workers Union
THORP	= Thermal Oxide Reprocessing Plant
TLD	= Thermoluminescent Dosimeter
TUC	= Trades Union Congress
UKAEA	= United Kingdom Atomic Energy Authority
UNSCEAR	= United Nations Scientific Committee on the Effects of Atomic Radiation
WHO	= World Health Organisation
WLM	= Working Level Month
YTV	= Yorkshire Television

References

H.J. Alder and A.M. Weinberg, An Approach to Setting Radiation Standards, *Health Physics*, Vol. 34, June 1978, pp.719-720.

G.H. Alcalay, The Aftermath of Bikini, *The Ecologist*, Vol.10, No.10, December 1980, pp.346-351.

Anon. *Rocky Flats Nuclear Bomb Factory*, 1977.

V.E. Archer, Geomagnetism, Cancer, Weather and Cosmic Radiation, *Health Physics*, Vol.34, March 1978, pp.237-247.

V.E. Archer, Cancer and Anencephalus in Man Associated with Background Radiation, in *Energy and Health* edited by N.E. Breslow and A.S. Whittemore, Society for Industrial and Applied Mathematics, 1979, pp.81-96.

V.E. Archer,, J.D. Gilliam and J.K. Wagoner, Respiratory Disease Mortality Among Uranium Miners, *Annals of the New York Academy of Sciences*, Vol.271, 1976, pp.280-293.

V.E. Archer, J.K. Wagoner and F.E. Lundin, Lung Cancer Among Uranium Miners in the United States, *Health Physics*, Vol.25, October 1973, pp.351-371.

A. Arends, Pledge on N-Test Probe Questioned, *The Financial Times*, 4th January, 1985, page 6.

Atom, Blix Stresses Importance of US Participation, *Atom,* No.315, January, 1983a, page 18.

Atom, Compensation in Nuclear Industry, *Atom*, No.318, April, 1983b, page 88.

Atom, London Dumping Convention, *Atom*, No. 318, April, 1983c, page 88.

Atom, Polonium Release in Windscale Fire, *Atom*, N0.320, June, 1983d, Page 140.

Atom, Nuclear Waste in Sea, *Atom*, No.327, January, 1984a, pp.27-28.

Atom, Nuclear Waste Disposal, *Atom*, No.327, January, 1984b, page 2
Atom, BNFL Payout, *Atom*, No.329, March, 1984c, pp.74-75.
Atom, Sellafield Discharges, *Atom*, No.330, April, 1984d, page 26.
Atom, Nuclear Waste, *Atom*, No.333, July, 1984e, pp.30-31.
Atom, Nuclear Waste, *Atom*, No.339, January, 1985a, pp.33-34.
Atom, Radioactive Waste from Hospitals, *Atom*, No.344, June, 1985b, pp.37-39.
Atom, BNFL Trial Ends, *Atom*, No.347, September, 1985c, page 16.
Atom, Solving the Problem that Nobody Wants, *Atom*, No.350, December, 1985d, pp.20-21.
Atom, BNFL Makes Compensation Payment, *Atom*, No.350, December, 1985e, page 16.
Atom, Radiation, *Atom*, No.352, February, 1986a, pp.46-47
Atom, Nuclear Waste, *Atom*, No.353, March, 1986b, page 41.
Atom, Nuclear Installations, *Atom*, No.355, May, 1986c, page 49.
J.C. Bailar, Mammography: A Contrary View, *Annuals of Internal Medicine*, Vol.84, No.1, January, 1976, pp.77-84.
J.C. Bailar, Radiation Hazards of X-Ray Mammography, in *Late Biological Effects of Ionizing Radiation,* International Atomic Energy Agency, Vienna, 1978, Vol.2, pp.251-261.
M.A. Barcinski, M.D.C. Abreu, J.C. de Almeida, J.M. Naya, L.G. Fonseca and L.E. Castro, Cytogenetic Investigation in a Brazilian Population Living in an Area of High Natural Radioactivity, *American Journal of Human Genetics,* Vol.27, No.6, November, 1975, pp.802-806.
D.L. Barlett and J.B. Steele, *Forevermore: Nuclear Waste in America,* W.W. Norton and Co., 1985.
F. Barnaby, The Controversy Over Low-Level Radiation, *Ambio*, Vol. 9, No.2, 1980, pp.74-80.
C.J. Barton, E. Roman, H.M. Ryder and A. Watson, Childhood Leukaemia in West Berskhire, *The Lancet*, No.8466, 30 November, 1985, pp.1248-1249.
J.W. Baum, Population Heterogeneity Hypothesis on Radiation Induced Cancer, *Health Physics*, Vol. 25, August 1973, pp.97-104.
K.F. Baverstock, D.G. Papworth and J. Vennart, Incidence of Myeloid Leukaemia in Lancashaire, *The Lancet*, 22 and 29 December 1979, pp.1362-1363.
J. A. Bean, P. Isacson, R. M. A. Hahne and J. Kohler, Drinking Water and Cancer Incidence in Iowa: II Radioactivity in Drinking Water, *American Journal of Epidemiology*, Vol.16, No.6, 1982, pp.924-932.
J.R. Beattie, R.F. Griffiths, G.D. Kaiser and G.H. Kinchin, The Environmental Impact of Radioactive Releases from Accidents in Nuclear Power Reactors, in *Nuclear Energy and the Environment* , edited by E.E. El-Hinnawi, Pergamon Press, Oxford, 1980, pp.95-138.
L.B. Beentjes, J.J. Broese, A.J. van der Kogel and A. van der Wielen, Age Dependence of the Risk of Radiation Induced Fatal Malignancies, *Health Physics*, Vol 38, February 1980, pp.239-241.
V. Beral, H. Inskip, P. Fraser, M. Booth, D. Coleman and G. Rose, Mortality of Employees of the United Kingdom Atomic Energy Authority, 1946-1979, *British Medical Journal*, Vol.291, No.6493, 17 August 1985, pp.440-447.

R. Bertell, X-Ray Exposure and Premature Ageing, *Journal of Surgical Oncology*, Vol.9, 1977, pp.379-391.
R. Bertell, Nuclear Crossroads, in *A Nuclear Ireland?*, edited by J.F. Carroll and P.K. Kelley, Irish TGWU, Dublin, 1978, pp.229-236.
R. Bertell, The Nuclear Worker and Ionising Radiation, *American Industrial Hygiene Association Journal*, Vol.40, May 1979, pp.395-401.
R. Bertell, Response by Dr. Rosalie Bertell, *Environmental Health Review*, Vo.26, No.2, June, 1982, pp.47-48.
R. Bertell, *Handbook for Estimating Health Effects from Exposure to Ionizing Radiation*, International Radiation Research and Training Institute, Regional Cancer Registry, Queen Elizabeth Medical Centre, Birmingham, 1984.
R. Bertell, *No Immediate Danger: Prognosis for a Radioactive Earth* , Womens Press, 1985.
BEIR, *The Effects on Populations of Exposure to Low Levels of Ionizing Radiation*, Report of the Advisory Committee on the Biological Effects of Ionizing Radiations, National Academy of Sciences, National Academy Press, 1980.
J.M. Birch, H.B. Marsden and R. Swindell, Acute Lymphoid Leukaemia of Childhood in North West England, *The Lancet*, No.8147, 20 October, 1979, pp.854-855.
J.F. Bithell and A.M.D. Stewart, Pre-Natal Irradiation and Childhood Malignancy: A Review of British Data from the Oxford Survey, *British Journal of Cancer*, Vol.31, 1975, pp.271-287.
D. Black, *Investigation of the Possible Increased Incidence of Cancer in West Cumbria*, H.M.S.O., 1984.
W.J. Blair and R.C. Thompson, Plutonium: Biomedical Research, *Science*, Vol.183, No.4126, 22 February 1974, pp.715-722.
H.M. Blalock, *Causal Inferences in Nonexperimental Research*, University of North Carolina Press, 1961.
J.W. Boag, J. Fielding, J.H. Humphrey, A. Jacobs, P. Lindop, J. Rotblat and J.A. Thompson, Cancer Following Nuclear Weapons Tests, *The Lancet*, No.8328, 9 April, 1983, page 815.
D. Bodanis, The Frying Squad, *The Guardian*, 4 November, 1982.
P.A. Boffey, J. Gofman and A. Tamplin: Harrassment Charges Against AEC, Livermore, *Science*, Vol.169, No.3948, 28 August 1970, pp.838-843.
J.D. Boice and C.E. Land, Adult Leukaemia Following Diagnostic X-Rays? (Review of Report by Bross, Ball and Falen on a Tri-State Leukaemia Survey), *American Journal of Public Health*, Vol.69, No.2, February 1979, pp.137-145.
J.A. Bonnell and G. Harte, Risk Associated with Occupational Exposure to Ionizing Radiation kept in Perspective, in *Late Biological Effects of Ionizing Radiation*, International Atomic Energy Agency, Vol.1, Vienna, 1978, pp.413-425.
C. Bowlt and J.R. Howe, Radioactive Iodine-125 and Iodine-129 Derived from Environmental Pollution in Members of the Public, *The Lancet*, No.8469/70, 21 and 28 December, 1985, page 1420.
J.T. Boyd, R. Doll, J.S. Faulds and J. Leiper, Cancer of the Lung in Iron Ore (Haematite) Miners, *British Journal of Industrial Medicine*, Vol.27, 1970, pp.97-105.

W.F. Brandom, A.D. Bloom, P.G. Archer, V.E. Archer, R.W. Bistline and G. Saccomanno, Somatic Cell Genetics of Uranium Miners and Plutonium Workers, in *Late Biological Effects of Ionizing Radiation* , International Atomic Energy Agency, Vol.1, Vienna, 1978, pp.507-518.
D. Brindle and D. Fishlock, Plan for N-Waste Dumping Inquiry, *The Financial Times*, 6 December, 1983.
British Nuclear Forum, Disposal of Radioactive Waste in the Deep Ocean, *British Nuclear Forum*, July 1983.
British Nuclear Fuels Limited, *Comments on the Report 'The Impact of Nuclear Waste Disposals to the Marine Environment', by the Political Ecology Research Group (RR-8)* , BNFL, Warrington, June, 1982, 13 pages.
British Nuclear Fuels Limited, *Annual Report and Accounts for the Year Ended 31 March 1984*, BNFL, Warrington, 1984.
I.D.J. Bross, *A 1980 Reassessment of the Health Hazards of Low-Level Ionizing Radiation,* Paper given at the University of Heidelberg, Germany, 29 October, 1979, 36 pages.
I.D.J. Bross, *Scientific Strategies to Save Your Life* , Marcel Dekker, 1981.
I.D.J. Bross, M. Ball and S. Falen, A Dosage Response Curve for the One Rad Range: Adult Risks from Diagnostic Radiation, *American Journal of Public Health*, Vol.69, No.2, February 1979, pp.130-136.
I.D.J. Bross, M. Ball, T. Rzepka and R.E. Laws, Preliminary Report on Radiation and Heart Disease, *Journal of Medicine*, Vol.9, No.1, 1978, pp.3-15.
I.D.J. Bross and N. Natarajan, Leukaemia from Low Level Radiation: Identification of Susceptible Children, *New England Journal of Medicine*, Vol.287, No.3, 20 July, 1972, pp.107-110.
I.D.J. Bross and N. Natarajan, Risk of Leukaemia in Susceptible Children Exposed to Preconception, In Utero and Postnatal Radiation, *Preventive Medicine*, Vol.3, 1974, pp.361-369.
I.D.J. Bross and N. Natarajan, Genetic Damage from Diagnostic Radiation, *Journal of the American Medical Association* , Vol.237, No.22, 30 May, 1977, pp.2399-2401.
I.D.J. Bross and N. Natarajan, Cumulative Genetic Damage in Children Exposed to Preconception and Intrauterine Radiation, *Investigative Radiology*, Vol.15, January/February 1980, pp.52-64.
J.M. Brown, Linearity versus Non-Linearity of Dose Response for Radiation Carcinogenesis, *Health Physics* , Vol.31, September, 1976, pp.231-245.
J.M. Brown, The Shape of the Dose-Response Curve for Radiation Carcinogenesis: Extrapolation to Low Doses, *Radiation Research* , Vol.71, 1977, pp.34-50.
P. Brown, Paradise Islanders Escape U.S. Radiation, *The Guardian*, 25 May, 1985, pages 1 and 30.
P. Brown and P. Johnson, Haughey Calls for Sellafield to be Closed, *The Guardian*, 30 July 1984.
P. Bunyard, *Nuclear Britain*, New English Library, 1981.
H. Caldicott, *Nuclear Madness*, Bantam Books, 1980.
G.G. Caldwell, D.B. Kelley and C.W. Heath, Leukaemia Among Participants in Military Manoeuvers at a Nuclear Bomb Blast, *Journal*

of the American Medical Association , Vol.244, No.14, 3 October, 1980, pp.1575-1578.

G.G. Caldwell, D.B. Kelley, M. Zack, H. Falk and C.W. Heath, Mortality and Cancer Frequency Among Military Nuclear Test (Smoky) Participants, 1957 through 1979, *Journal of the American Medical Association*, Vol.250, No.5, 5 August, 1983, pp.620-624.

R.S. Cambray and J.D. Eakins, Pu, Am^{241} and Cs^{137} in Soil in West Cumbria and a Maritime Effect, *Nature*, Vol.300, No.5887, 4 November, 1982, pp.46-48.

Canadian Labour Congress, *Submission of the Canadian Labour Congress to the Atomic Energy Control Board on Proposed Revisions to Regulations Under the Atomic Control Act* , Canadian Labour Congress, January 1984, 45 pages.

M.W. Charles and P.J. Lindop, Risk Assessment Without the Bombs, *Journal of the Society for Radiological Protection* , Vol.1, No.3, 1981, pp.15-19.

M.W. Charles, P.J. Lindop and A.J. Mill, Pragmatic Evaluation of Repercussions for Radiological Protection of Recent Revisions in Japanese A-Bomb Dosimetry, in *Biological Effects of Low-Level Radiation*, Proceedings of a Symposium, Venice, April, 1983, International Atomic Energy Agency, Vienna, 1983, pp.61-75.

S. Chinn, *The Relation of the Rocky Flats Plant and Other Factors to 1969-1971 Cancer Incidence in the Denver Area* , Fairfield and Woods, September 1981, 79 pages.

P. Chorlton, Confusion Persists Over Radiation Effects, *The Guardian*, 26 February, 1983.

M.J. Clark and A.B. Fleishman, The Valuation of Detriment, in *Radiation Protection Optimization,* edited by A. Oudiz, H. Ebert, G. Uzzan and H. Eriskat, Pergamon Press, 1981a, pp.75-84.

M.J. Clark and A.B. Fleishman, The Implications of Cost Benefit Analysis for Optimization, in *Radiation Protection Optimization* edited by A. Oudiz, H. Ebert,. G. Uzzan and H. Eriskat, Pergamon Press, 1981b, pp.85-95.

M.J. Clark, A.B. Fleishman and G.A.M. Webb, *Optimization of the Radiological Protection of the Public* , NRPB-R120, HMSO, July 1981c.

V.P. Coffey and M.B. Hillary, A Review of Down Syndrome in the Eastern Health Board Region of Ireland - 1979 and 1980, *Irish Medical Journal*, Vol. 75, No.8, August, 1982, pp.276-278.

B.L. Cohen, Health Effects of Radon Emissions from Uranium Mill Tailings, *Health Physics*, Vol.42, No.5, May, 1982, pp.695-702.

J.J. Cohen, On Determining the Cost of Radiation Exposure to Populations for Purposes of Cost-Benefit Analyses, *Health Physics*, Vol.25, November, 1973, pp.527-528.

C. Comar, Risk: A Pragmatic De Minimis Approach, *Science*, Vol.203, No.4378, 26 January, 1979, page 320.

D.D. Comey, The Legacy of Uranium Tailings, *The Bulletin of the Atomic Scientists*, September, 1975, pp.43-45.

Comptroller General, *Problems in Assessing the Cancer Risk of Low-Level Ionizing Radiation Exposure*, Report to the Congress of the United States, General Accounting Office, January 1981, 85 pages.

R.A. Conard, Summary of Thyroid Findings in Marshallese 22 Years After Exposure to Radioactive Fallout, in *Radiation-Associated Thyroid Carcinoma,* edited by L.J. DeGroot, Grune and Stratton, New York, 1977, pp.241-257.
C. Costa-Ribeiro, M.A. Barcinski, N. Figueiredo, E. Penna Franca, N. Lobao and H. Krieger, Radiobiological Aspects and Radiation Levels Associated with the Milling of Monazite Sand, *Health Physics,* Vol.28, March 1975, pp.225-231.
W.M. Court Brown, R. Doll, F.W. Spiers, B.J. Duffy and M.J. McHugh, Geographical Variation in Leukaemia Mortality in Relation to Background Radiation and Other Factors, *British Medical Journal* , 11 June, 1960, pp.1753-1759.
A.W. Craft and S. Openshaw, Childhood Cancer in West Cumbria, *The Lancet*, No.8425, 16 February, 1985, page 403.
M.J. Crick and G.S. Linsley, *An Assessment of the Radiological Impact of the Windscale Reactor Fire , October, 1957*, National Radiological Protection Board, November 1982, NRPB - R135.
J. Cutler, The Perils that Lurk by the Sea, *The Guardian*, 23 July, 1984, page 15.
J. Cutler, Checking the Figures, *The New Statesman* , Vol.109, No.2809, 18 January, 1985, pp.10-11.
Daily Telegraph, RAF Veteran, 84, Tells of Flight in Atom Cloud, *The Daily Telegraph*, 12 February, 1985, page 3.
T. Dalyell, Political Front-Door Steps, *New Scientist*, 25 March, 1982, page 303.
N. Davies, Veterans of Atomic Tests Begin Fight for Ministry Compensation, *The Guardian*, 5 April, 1983, page 3.
J.P. Day and J.E. Cross, Americium-241 From the Decay of Plutonium-241 in the Irish Sea, *Nature*, Vol.292, No.5818, 2 and 8 July, 1981, pp.43-45.
M.H. Degroot, Statistical Studies of the Effect of Low Level Radiation from Nuclear Reactors on Human Health, in *Proceedings of the Sixth Berkeley Symposium on Mathematical Statistics and Probability*, edited by L.M. Le Cam, J. Neyman and E. Scott, University of California Press, 1972, Vol.6, pp.224-234.
D. Dickson, Home Insulation May Increase Radiation Hazard, *Nature*, Vol.276, No.5687, 30 November, 1978, page 431.
D. Dickson, Sparks Continue to Fly in Low-Level Radiation Row, *Nature*, Vol.280, No.5719, 19 July, 1979, pp.180-181.
P.W. Durbin, Plutonium in Man: A New Look at the Old Data, in *The Radiobiology of Plutonium,* edited by B.J. Stover and W.S.S. Jee, The J.W. Press, 1972, pp.469-530.
J.D. Eakins and R.S. Cambray, *Studies of Environmental Radioactivity in Cumbria, Part 6: The Chronology of Discharges of Caesium-137, Plutonium and Americium-241 from BNFL, Sellafield as Recorded in Lake Sediments*, AERE-R-11181, HMSO, April, 1985.
J.D. Eakins, A.E. Lally, P.J. Burton, D.R. Kilworth and F.A. Pratley, *Studies of the Environmental Radioactivity in Cumbria: Part 5. The Magnitude and Mechanism of Enrichment of Sea Spray with Actinides in West Cumbria*, AERE-R 10127, HMSO, March, 1982.
Editorial, Nuclear Power, *The Lancet*, 12 May, 1979, page 1023.
R.E. Ellis, Proof of Evidence, in *Planning and Plutonium*, Town and County Planning Association, 1978, pp.49-67.

Environment Committee of the House of Commons, *Radioactive Waste*, First Report from the Environment Committee, Vol.1, HMSO, 1986.
H.J. Evans, K.E. Buckton, G.E. Hamilton and A. Carothers, Radiation Induced Chromosome Aberrations in Nuclear Dockyard Workers, *Nature*, Vol.277, No.5697, 15 February, 1979, pp.531-534.
Financial Times, N-Case Retrial, *The Financial Times*, 6 August, 1982, page 1.
Financial Times, Scientists See No Threat to Nuclear Workers, *The Financial Times*, 8 January, 1983a.
Financial Times, Sizewell B's Cost Estimate Too Low, *The Financial Times*, 27 May, 1983b, page 8.
Financial Times, Sizewell B, *The Financial Times*, 28 September, 1983c.
Financial Times, Plutonium at Test Site, *The Financial Times*, 1 June 1984, page 1.
D. Fishlock, The 'Watchdog' Over Windscale: A Response to the Cancer Claim, *The Financial Times*, 4 November, 1983.
D. Fishlock, Compensation Call for Neighbours of Nuclear Stations, *The Financial Times*, 4 April, 1984a.
D. Fishlock, Japan 'Pays N-Plant Indemnities', *The Financial Times*, 8 September, 1984b.
A.B. Fleishman, *The Significance of Small Doses of Radiation to Members of the Public*, NRPB-R-175, HMSO, March 1985.
F.A. Fry, N. Green, N.J. Dodd and D.J. Hammond, *Radionuclides in House Dust*, NRPB-R181, HMSO, April, 1985.
R.M. Fry and J.E. Cook, Health Effects of Radon-222 from Uranium Mining: Comment, *Search*, Vol.7, No.8, August 1976, pp.351-353.
P.S. Furcinitti and P. Todd, Gamma Rays: Further Evidence for Lack of a Threshold Dose for Lethality to Human Cells, *Science*, Vol.206, No.4417, 26 October, 1979, pp.475-477.
M.J. Gardner, Childhood Cancer in West Cumbria, *The Lancet*, No.8425, 16 February, 1985, pp.403-404.
C.G. Geary, R.T. Benn and I. Leck, Incidence of Myeloid Leukaemia in Lancashire, *The Lancet*, No.8142, 15 December 1979, pp.549-55.
GMBATU, *Evidence to the Sizewell Inquiry*, General Municipal, Boilermakers and Allied Trades Union, 1983.
R. Gibson, S. Graham, A. Lilienfeld, L. Schuman, J.E. Dowd and M.L. Levin, Irradiation in the Epidemiology of Leukaemia Among Adults, *Journal of the National Cancer Institute*, Vol.48, No.2, February, 1972, pp.301-311.
E.S. Gilbert, An Evaluation of Several Methods for Assessing the Effects of Occupational Exposure to Radiation, *Biometrics*, Vol.39, March, 1983, pp.161-171.
J.W. Gofman, The Plutonium Controversy, *Journal of the American Medical Association*, Vol.236, No.3, 19 July, 1976a, pp.284-286.
J.W. Gofman, An Examination of Some of the Criticisms of Those in Opposition to the Nuclear Industry, in *The Environmental Impact of Nuclear Power Plants*, edited by R.A. Karam and K.Z. Morgan, Pergamon Press, 1976b, pp.532-546.
J.W. Gofman, The Question of Radiation Causation of Cancer in Hanford Workers, *Health Physics*, Vol.37, November, 1979, pp.617-639.
J.W. Gofman, *Radiation and Human Health*, Pantheon Books, updated and abridged edition, 1983.

J.W. Gofman and A.R. Tamplin, Epidemiologic Studies of Carcinogenesis by Ionizing Radiation, in *Proceedings of the Sixth Berkeley Symposium on Mathematical Statistics and Probability*, edited by L.M. Le Cam, J. Neyman and E. Scott, University of California Press, 1972, Vol.6, pp.235-277.

J.W. Gofman and A.R. Tamplin, *Poisoned Power*, Second edition, Rodale Press, 1979.

D.W. Gorst and C. Atkinson, High Incidence of Adult Acute Leukaemia in North-West Lancashire, *The Lancet*, No.8416, 15 December, 1984, pp.1397-1398.

S.G. Goss, Sizes of Population Needed to Detect an Increase in Disease Risk when the Level of Risk in the Exposed and the Controls are Specified: Examples from Cancer Induced by Ionizing Radiation, *Health Physics*, Vol.29, November, 1975, pp.715-721,

S.G. Goss, Integrity and the NRPB, *New Scientist*, Vol.73, No.1034, 13 January, 1977, page 103.

S. Graham, M.L. Levin, A.M. Lilienfeld, J.E. Dowd, L.M. Schuman, R. Gibson, L.H. Hempelmann and P. Gerhardt, Methodological Problems and Design of the Tri-state Leukaemia Survey, *Annals of the New York Academy of Science*, Vol.107, 1963, pp.557-569.

K. Grossman, *Cover Up: What You Are Not Supposed to Know About Nuclear Power*, The Permanent Press, 1980.

Guardian, Radiation Protection Watchdog Attacked, *The Guardian*, 19 September, 1984a.

Guardian, Windscale Verdict, *The Guardian*, 20 October, 1984b.

I. Hamilton, IoM Calls for Inspection of Sellafield, *The Financial Times*, 18 July, 1986.

T. Handley, Concorde Evidence on Cancer Scare, *Reading Chronicle*, 24 January, 1986.

J.D. Harrison, The Gut Uptake of Actinides, *Radiological Protection Bulletin*, No.41, 1981, pp.14-19.

M.A. Heasman, I.W. Kemp, J.D. Urquhart and R. Black, Childhood Leukaemia in Northern Scotland, *The Lancet*, No.8475, 1 February, 1986, page 266.

G.W.F. Hegel, *Phenomenology of Spirit*, Translated by A.V. Miller, Clarendon Press, (1807), 1977.

C.R. Hemming and R.H. Clarke, *A Review of Environmental Radiation Protection Standards*, NRPB-R168, HMSO, October 1984.

L.H. Hempelmann, J.W. Hall, M. Phillips, R.A.Cooper and W.R. Ames, Neoplasms in Persons Treated with X-rays in Infancy: Fourth Survey in 20 Years, *Journal of the National Cancer Institute*, Vol.55, No.3, September, 1975, pp.519-530.

N. Heneson, Heads it Causes Cancer, Tails it Doesn't, *New Scientist*, Vol.98, No.1356, 5 May, 1983, page 275.

R. Herbert, The Day the Reactor Caught Fire, *New Scientist*, Vol.96, No.1327, 14 October, 1982, pp.84-87.

R. Hesketh, The Curious Business of British Nuclear Fuel's Plutonium Sting, *The Guardian*, 8 November, 1984.

High Background Radiation Research Group, Health Survey in High Background Radiation Areas in China, *Science*, Vol.209, No.4451, 22 August, 1980, pp.877-880.

R. M. Holford, The Relation Between Juvenile Cancer and Obstetric Radiography, *Health Physics*, Vol.28, February, 1975, pp.153-156.

F. G. T. Holliday, *Report of the Independent Review of Disposal of Radioactive Waste in the North Sea,* HMSO, 1984.

D. Housego, Fabius Admits that France Sank Boat, *The Financial Times*, 23rd September, 1985, page 1.

T. W. Hu and K. S. Slaysman, Health-Related Economic Costs of the Three-Mile Island Accident, *Socio-Economic Planning Sciences* , Vol.18, No.3, 1984, pp.183-193.

ICRP, *General Principles of Monitoring for Radiation Protection of Workers*, ICRP Publication No.12, Pergamon Press, 1969.

ICRP, Recommendations of the International Commission on Radiological Protection, ICRP Publication No.26, *Annals of the ICRP* , Vol.1, No.3, Pergamon Press, 1977.

A.P. Jacobson, P.A. Plato and N.A. Frigeiro, The Role of Natural Radiation in Human Leukemogenesis, *American Journal of Public Health,* Vol.66, No.1, January 1976, pp.31-37.

C.J. Johnson, *Epidemiological Evaluation of Cancer Incidence Rates for the Period 1969-71 in Areas of Census Tracts with Measured Concentration of Plutonium Soil Contamination Downwind from the Rocky Flats Plant.* A Report to the Jefferson County Board of Health, The Colorado Board of Health and the National Cancer Institute, February 1979.

C.J. Johnson, Evaluation of Cancer Incidence for Anglos in the Period 1969-71 in Areas of Census Tracts with Measured Concentrations of Plutonium Soil Contamination Downwind from the Rocky Flats Plant in the Denver Standard Metropolitan Statistical Area, in *Proceedings of the Fifth International Congress of the International Radiation Protection Association*, March 1980, pp.365-368.

C.J. Johnson, Cancer Incidence in an Area Contaminated with Radionuclides Near a Nuclear Installation, *Ambio*, Vol.10, No.4, 1981, pp.176-182.

C.J. Johnson, Cancer Incidence in an Area of Radioactive Fallout Downwind from the Nevada Test Site, *Journal of the American Medical Association* , Vol.251, No.2, 13 January, 1984, pp.230-236.

C.J. Johnson, R.R. Tidball and R.C. Severson, Plutonium Hazard in Respirable Dust on the Surface of Soil, *Science*, Vol.193, No.4252, 6 August, 1976, pp.448-490.

R.R. Jones, Leukaemia and Sellafield, *The Lancet,* No.8400, 25 August, 1984, pp.467-468.

M.W. Jones-Lee, *The Value of Human Life: An Economic Analysis* , Martin Robertson, 1976.

M.W. Jones-Lee, The Value of Non-Marginal Changes in Physical Risk, in *Microeconomic Analysis* edited by D. Currie, D. Peel and W. Peters, Croom Helm, 1981, pp.233-268.

G.W. Kneale, T.F. Mancuso and A.M.D. Stewart, Hanford Radiation Study III: A Cohort Study of the Cancer Risks from Radiation to Workers at Hanford (1944-77 deaths) by the Method of Regression Models in Life Tables, *British Journal of Industrial Medicine* , Vol.38, 1981, pp.156-166.

G.W. Kneale and A.M.D. Stewart, Pre Cancer and Liability to other Diseases, *British Journal of Cancer*, Vol.37, 1978, pp.448-457.

G.W. Kneale, A.M.D. Stewart and T.F. Mancuso, Re-Analysis of Data Relating to the Hanford Study of the Cancer Risks of Radiation

Workers, in *Late Biological Effects of Ionizing Radiation* , IAEA, Vienna, 1978, Vol.1, pp.387-412.
E.G. Knox, T. Sorahan and A.M.D. Stewart, Cancer Following Nuclear Weapons Tests, *The Lancet*, No.8328, 19 April, 1983, page 815.
N. Kochupillai, I.C. Verma, M.S. Grewal and V. Ramalingaswami, Down's Syndrome and Related Abnormalities in an Area of High Background Radiation in Coastal Kerala, *Nature*, Vol.262, No.5561, 1 July, 1976, pp.60-61.
C.E. Land, Estimating Cancer Risks from Low Doses of Ionizing Radiation, *Science*, Vol.209, No.4462, 12 September, 1980, pp.1197-1203.
E. Landau, Health Effects of Low Dose Radiation. Problems of Assessment, *International Journal of Environmental Studies*, Vol.6, 1974, pp.51-57.
R.E. Lapp, *The Radiation Controversy* , Reddy Communications Inc., 1979.
P.R. Larsen, R.A. Conard, K. Knudsen, J. Robbins, J. Wolff, J.E. Rall and B.Dobyns, Thyroid Hypofunction Appearing as a Delayed Manifestation of Accidental Exposure to Radioactive Fall-out in a Marshallese Population, in *Late Biological Effects of Ionizing Radiation*, International Atomic Energy Agency, Vienna, 1978, Vol.2, pp.101-115.
R.P. Larsen and R.D. Oldham, Plutonium in Drinking Water: Effects of Chlorination on its Maximum Permissible Concentration, *Science*, Vol.201, 15 September, 1978, pp.1008-1009.
G. Lean, Radon: Second Largest Cause of Cancer, *The Observer* , July, 1986.
G. Lean and W. Patterson, The Radioactive Sea, *The Observer* , 20 May, 1984, pp.16-23.
I. Leck, R.T. Benn and A. Smith, Changes in Myeloid Leukaemia Incidence, *The Lancet*, No.8197, 4 October, 1980, pp.749-750.
D. Leigh and P. Lashmar, MoD Funds Own A-Test History, *The Observer*, 23 June, 1985.
R. Lewis, Shippingport - The Killer Reactor? *New Scientist*, Vol.59, No.862, 6 September, 1973, pp.552-553
P.J. Lindop, *Radiation Hazards, Areas of Uncertainty*, Mimeo, 1978?
W.E. Loewe and E. Mendelsohn, Revised Dose Estimates at Hiroshima and Nagasaki, *Health Physics*, Vol.41, No.4, October, 1981, pp.663-666.
A.B. Lovins and W.C. Paterson, Plutonium Particles: Some Like Them Hot, *Nature*, Vol.254, No.5498, 27 March, 1975, pp.278-280.
F.E. Lundin, V.E. Archer and J.K. Wagoner, An Exposure-Time-Response Model for Lung Cancer Mortality in Uranium Miners: Effects of Radiation Exposure, Age, and Cigarette Smoking, in *Energy and Health*, edited by N.E. Breslow and A.S. Whittemore, Society for Industrial and Applied Mathematics, 1979, pp.243-264.
J.L. Lyon, M.R. Klauber, J.W. Gardner and K.S. Udall, Childhood Leukaemias Associated with Fallout from Nuclear Testing, *New England Journal of Medicine*, Vol.300, No.8, 22 February, 1979, pp.397-402.
C. MacDougall, Concern Grows in China Over Safety of Nuclear Pragramme, *The Financial Times*, 23 May, 1986, page 3.
S. Macgill, Living With Sellafield, *Atom*, No.355, May, 1986, pp.2-4.

D.O. MacKenzie, Mururoa Sinks Under a 'Force de Frappe', *New Scientist*, Vol.94, No.1301, 15 April, 1982, pp.170-171.
T.M. Mack, M.C. Pike and J.T. Casagrande, Epidemiologic Methods for Human Risk Assessment, in *Origins of Human Cancer*, edited by H.H. Hiatt, J.D. Watson and J.A. Winsten, Cold Spring Harbour Laboratory, Vol.C, 1977, pp.1749-1763.
J.R. McClelland, J. Fitch and W.J.A. Jonas, *The Report of the Royal Commission into British Nuclear Tests in Australia*, Conclusions and Recommendations, Australian Government Publishing Service, Canberra, 1985.
J.L. McClenahan, Wasted X-Rays, *Pennsylvania Medicine* , Vol.72, November, 1969, pp.107-108. Reprinted in *Radiology*, Vol.96, August, 1970, pp.453-456.
G.K. Macleod, Some Public Health Lessons from Three Mile Island: A Case Study in Chaos, *Ambio*, Vol.10, No.1, 1981, pp.18-23.
B. MacMahon, Prenatal X-Ray Exposure and Childhood Cancer, *Journal of the National Cancer Institute*, Vol.28, No.5, May 1962, pp.1173-1191.
T.F. Mancuso, A.M.D. Stewart and G.W. Kneale, Radiation Exposure of Hanford Workers Dying from Cancer and Other Causes, *Health Physics*, Vol.33, No.5, November 1977, pp.369-385.
E. Marshall, New A-Bomb Studies Alter Radiation Estimates, *Science*, Vol.212, No.4497, 22 May, 1981a, pp.900-903.
E. Marshall, New A-Bomb Data Shown to Radiation Experts, *Science*, Vol.212, No.4501, 19 June, 1981b, pp.1364-1365.
E. Marshall, Japanese A-Bomb Data will be Revised, *Science*, Vol.214, No.4516, 2 October, 1981c, pp.31-32.
E.A. Martell, Radioactivity of Tobacco Trichomes and Insoluble Cigarette Smoke Particles, *Nature*, Vol.249, 17 May, 1974, pp.215-217.
E.A. Martell, Tobacco Radioactivity and Cancer in Smokers, *American Scientist*, Vol.63, No.4, July/August, 1975, pp.404-412.
J.L. Marx, Low-Level Radiation: Just How Bad Is It? *Science*, Vol.204, No.4389, 13 April, 1979, pp.160-164.
J.A. Matthews, A Critique of the Internal Logic of ICRP Risk Factor Derivation, *Annals of Occupational Hygiene* , Vol.27, No.3, 1983, pp.291-300.
C.W. Mays, Cancer Induction in Man from Internal Radioactivity, *Health Physics*, Vol.25, December 1973, pp.585-592.
C.W. Mays, H. Spiess and A. Gerspach, Skelatel Effects Following ^{224}Ra Injections into Humans, *Health Physics*, Vol.35, July 1978, pp.83-90.
A.T. Meadows, D.J. Massari, J. Fergusson, J. Gordon, P. Littman and K. Moss. Declines in IQ Scores and Cognitive Dysfunctions in Children with Acute Lymphocytic Leukaemia Treated with Cranial Irradiation, *The Lancet*, No.8254, 7 November, 1981, pp.1015-1018.
Z.A. Medvedev, *Nuclear Disaster in the Urals*, Vintage Books, 1980.
R. Milliken, Fallout of a Cloud of Secrecy, *The New Statesman* , Vol.108, No.2802, 30 November, 1984, pp.19-20.
R. Milne, Government Fears Tritium Pollution Over Thames Valley, *New Scientist*, Vol.108, No.1483, 21 November, 1985, p.20.

B. Modan, D. Baidatz, H. Mart, R.Steinitz and S.G. Levin, Radiation Induced Head and Neck Tumors, *The Lancet*, No.7852, 23 February, 1974, pp.277-279.

B. Modan, E. Ron and A. Werner, Thyroid Neoplasms in a Population Irradiated for Scalp Tinea in Childhood, in *Radiation-Associated Thyroid Carcinoma*, edited by L.J. DeGroot, Grune and Stratton, New York, 1977a, pp.449-457.

B. Modan, E. Ron and A. Werner, Thyroid Cancer Following Scalp Irradiation, *Radiology*, Vol.123, June, 1977b, pp.741-744.

R.H. Mole, Antenatal Irradiation and Childhood Cancer: Causation or Coincidence? *British Journal of Cancer*, Vol.30, 1974, pp.199-208.

R.H. Mole, Man, Rems and Risk, in *Radiation Protection Measurement - Philosophy and Implementation*, edited by P. Recht and J.R.A. Laxey, Commission of the European Communites; 1975a, pp.23-36.

R.H. Mole, Ionizing Radiation as a Carcinogen: Practical Questions and Academic Pursuits, *British Journal of Radiology* , Vol.48, No.567, March 1975b, pp.157-169.

R.H. Mole, The Radiological Significance of the Studies with Ra-224 and Thorotrast, *Health Physics*, Vol.35, July, 1978, pp.167-174.

R.H. Mole, Radiation Effects on Pre-Natal Development and their Radiological Significance, *British Journal of Radiology* , Vol.52, No.614, February, 1979, pp.89-101.

R.H.Mole, Effects of Ionizing Radiation, in *Trauma and After*, edited by R. Porter, J. Price and R.Read, Pitman Medical, 1981, pp.54-72.

J.R. Morgan, A History of Pitchblende, *Atom*, Vol.329, March, 1984, pp.63-68.

K.Z. Morgan, The Need for Radiation Protection, *Radiologic Technology*, Vol.44, No.6, 1973, pp.385-395.

K.Z. Morgan, Reducing Medical Exposure to Ionizing Radiation, *American Industrial Hygiene Association Journal* , Vol.36, May, 1975a, pp.358-368.

K.Z. Morgan, Suggested Reduction of Permissible Exposure to Plutonium and Other Transuranium Elements, *American Industrial Hygiene Association Journal*, Vol.36, August, 1975b, pp.567-575.

K.Z. Morgan, Cancer and Low Level Ionizing Radiation, *The Bulletin of Atomic Scientists*, Vol.34, September, 1978, pp.30-41.

K.Z. Morgan, How Dangerous is Low-Level Radiation? *New Scientist*, Vol.82, No.1149, 5 April, 1979a, pp.18-21.

K.Z. Morgan, *The Non-Threshold Dose/Effect Relationship* . Paper given to the Academy Forum of the National Academy of Science, Washington, D.C., 27 September, 1979b, 15 pages.

K.Z. Morgan, *Appreciation of Risks of Low Level Radiation vs Nuclear Energy*, Mimeo, January, 1980, 15 pages.

P.W. Mummery, H. Howells and A. Scriven, The Environmental Impact of Reprocessing, in *Nuclear Energy and the Environment* , edited by E.E. El-Hinnawi, Pergamon Press, Oxford, 1980, pp.139-167,

T. Najarian, The Controversy Over the Health Effects of Radiation, *Technology Review*, November, 1978, pp.74-82.

T. Najarian and T. Colton, Mortality from Leukaemia and Cancer in Shipyard Nuclear Workers, *The Lancet*, No.8072, 13 May, 1978, pp.1018-1020.

N. Natarajan and I.D.J. Bross, Preconception Radiation and Leukaemia, *Journal of Medicine*, Vol.4, 1973, pp.276-281.
National Radiological Protection Board, *The Application of Cost-Benefit Analysis to the Radiological Protection of the Public: A Consultative Document*, HMSO, March 1980, 75 pages.
National Radiological Protection Board, *Living with Radiation*, NRPB, HMSO, Second Edition, 1981.
K. Neumeister, Findings in Children After Radiation Exposure in Utero From X-Ray Examination of Mothers, in *Late Bilogical Effects of Ionizing Radiation*, International Atomic Energy Agency, Vol.1, Vienna, 1978, pp.119-134.
New Scientist, The Wasteful Truth About the Soviet Nuclear Disaster, *New Scientist*, Vol.85, No.1189, 10 January, 1980, p.61.
New Scientist, A-Bomb Islanders Take the U.S. to Court, *New Scientist*, Vol.89, 26 March, 1981, page 793.
New Scientist, Defence Chiefs Juggle Figures on Bomb Tests, *New Scientist*, Vol.98, No.1353, 14 April, 1983, p.62.
NIREX, *Deep Ocean Disposal of Low Level Radioactive Waste* , NIREX, 1983.
Y. Nishiwaki, Bikini Ash, *Atomic Scientists Journal* , Vol.4, 1954, pp.97-109.
Nuclear Power Information Group, *The Facts About Nuclear Energy* , UKAEA, July, 1983.
Office of Population Censuses and Surveys, *Cancer Statistics: Incidence, Survival and Mortality in England and Wales*, Studies on Medical and Population Subjects, No.43, HMSO, 1981.
M.C. O'Riordan and G.J. Hunt, *Radioactive Fluorescers in Dental Porcelains*, NRPB-R25, HMSO, May 1974.
C. Panati and M. Hudson, *The Silent Intruder: Surviving the Radiation Age*, Pan Books, 1982.
R.J. Parker, *The Windscale Inquiry: Report by the Hon. Justice Parker* , Vol.1, Report and Annexes, HMSO, 1978.
D.H. Peirson, R.S. Cambray, P.A. Cawse, J.D. Eakins and N.J. Pattenden, Environmental Radioactivity in Cumbria, *Nature*, Vol.300, No.5887, 4 November, 1982, pp.27-31.
N.J. Petersen, L.D. Samuels, H.F. Lucas and S.P. Abrahams, An Epidemiologic Approach to Low-Level Radium-226 Exposure, *Public Health Reports*, Vol.81, No.9, September, 1966, pp.805-814.
A. Petkau, Effect of $^{22}Na^{+}$ on a Phospholipid Membrane, *Health Physics*, Vol.22, March, 1972, pp.239-244.
B.Z. Pilch, C.R. Kahn, A.S. Ketcham and D. Henson, Thyroid Cancer After Radioactive Iodine Diagnostic Procedures in Childhood, *Paediatrics*, Vol.51, No.5, May, 1973, pp.898-902.
J. Pincent and L. Masse, Natural Radiation and Cancer Mortality in Several Areas of North Brittany, *International Journal of Epidemiology*, Vol.4, No.4, 1975, pp.311-316.
F.A.E. Pirani, No More Need for Nuclear Tests, *The Financial Times*, 8 January, 1986.
E.E. Pochin, Malignancies Following Low Radiation Exposures in Man, *The British Journal of Radiology*, Vo.49, July 1976a, pp.577-579.

E.E. Pochin, Problems Involved in Detecting Increased Malignancy Rates in Areas of High Natural Radiation Background, *Health Physics*, Vol.31, August, 1976b, pp.148-151.
E.E. Pochin, Assumption of Linearity in Dose-Effect Relationships, *Environmental Health Perspectives*, Vol.22, February 1978a, pp.103-105.
E.E. Pochin, *Report on Investigation into Radiological Health and Safety at the Ministry of Defence (Procurement Executive) Atomic Weapons Research Establishment, Aldermaston*, Harwell, 1978b.
E.E. Pochin, *Nuclear Radiation: Risks and Benefits*, Clarendon Press, Oxford, 1983.
R.O. Pohl, Health Effects of Radon-222 from Uranium Mining, *Search*, Vol.7, No.8, August 1976, pp.345-350.
J. Pohl-Ruling and P.Fischer, The Dose Effect Relationship of Chromosome Aberrations to and Irradiation in a Population Subjected to an Increased Burden of Natural Radioactivity, *Radiation Research*, Vol.80, 1979, pp.61-81.
J. Pohl-Ruling, P. Fischer and E. Pohl, Chromosome Aberrations in Peripheral Blood Lymphocytes Dependent on Various Dose Levels of Natural Radioactivity, in *Biological and Environmental Effects of Low Level Radiation*, International Atomic Energy Agency, Vol.2, Vienna, 1976, pp.317-324.
A.P. Polednak, A.F. Stehney and R.E. Rowland, Mortality Among Women First Employed Before 1930 in the US Radium Dial-Painting Industry, *American Journal of Epidemiology*, Vol.107, No.3, March, 1978, pp.179-195.
Political Ecology Research Group, *An Investigation into the Incidence of Cancer in the Areas Affected by Discharges from the Nuclear Fuels Reprocessing Plant at Windscale, Cumbria*, Research Report RR-6, PERG, December 1980.
A. Pomiankowski, Cancer Incidence at Sellafield, *Nature*, Vol.311, No.5982, 13 December, 1984, p.100.
D.S. Popplewell, G.J. Ham, T.E. Savory and W.R. Bradford, Radionuclide Concentrations in Some Cattle and Sheep from West Cumbria, *Radiological Protection Bulletin*, 1981, pp.20-23.
E.P. Radford, Human Health Effects of Low Doses of Ionizing Radiation: The BEIR III Controversy, *Radiation Research*, Vol.84, 1980, pp.369-394.
E.P. Radford, Scientific Controversy in the Public Domain, *Technology Review*, November/December, 1981a, pp.74-75.
E.P. Radford, *Statement Concerning Proposed Federal Radiation Protection Guidance for Occupational Exposures*, Hearings of US Environmental Protection Agency Office of Radiation Programs, 1981?b, 23 pages.
E.P. Radford, Epidemiology of Radiation-Induced Cancer, *Environmental Health Perspectives*, Vol.52, October, 1983, pp.45-50.
Radioactive Waste Management Advisory Committee, *Fifth Annual Report*, HMSO, June 1984.
R. Rashke, *The Killing of Karen Silkwood: The Story Behind the Kerr-McGhee Plutonium Case*, Penguin Books, 1982.
C. Reed, Atomic Test Town Fights to Prove Why so Many Had to Die of Cancer, *The Guardian*, 6 March, 1984.

S.D. Reid, T.M. Hall, C.M.B. Reid and P.J. Hull, Cluster of Myeloid Leukaemia in Lytham St. Annes, *The Lancet*, No.8142, 15 September, 1979, p.579.
J.S. Reissland and S.C. Darby, Comments on the Increased Cancer Incidence Reported in Denver, USA, *Radiological Protection Bulletin*, No.36, September, 1980, pp.16-22.
J.S. Reissland and V. Harries, A Scale for Measuring Risks, *New Scientist*, Vol.83, No.1172, 13 September 1979, pp.809-811.
D. Richings, Radiation Risks, Limits and ICRP, *New Scientist*, Vol.82, No.1152, 26 April, 1979, pp.278-280.
R.A. Rinsky, R.D. Zumwalde, R.J. Waxweiler, W.E. Murray, P.J. Bierbaum, P.J. Landrigan, M. Terpilak and C. Cox, Cancer Mortality at a Naval Nuclear Shipyard, *The Lancet*, No.8214, 31 January, 1981, pp.231-235.
L.E.J. Roberts, *Nuclear Power and Public Responsibility* , Cambridge University Press, 1984.
D. Rose and R. Norton-Taylor, £120,000 Paid to Family of Dead Sellafield Worker, *The Guardian*, 27 March, 1985, p.3.
J. Rotblat, The Puzzle of Absent Effects, *New Scientist*, Vol.75, No.1066, 25 August, 1977, pp.475-476.
J. Rotblat, The Risks for Radiation Workers, *The Bulletin of the Atomic Scientists*, Vol.34, September 1978, pp.41-46.
J. Rotblat, Radiation Hazards in Fission Fuel Cycles, in *The Hazards of the International Energy Crisis*, edited by D. Carlton and C. Schaerf, Macmillan, 1982, pp.107-126.
Royal College of Physicians, *Health or Smoking? Follow-up Report of the Royal College of Physicians*, Pitman, 1983.
Royal Commission on Environmental Pollution, *Sixth Report: Nuclear Power and the Environment*, HMSO, 1976, Cmnd.6618.
C. Ryle, J. Garrison and A.Webb, *Radiation: Your Health at Risk*, Radiation and Health Information Service, March 1980, 32 pages.
T.H. Saffer and O.E. Kelly, *Countdown Zero: GI Victims of US Atomic Testing*, Penguin Books, 1983.
J.M. Samet, D.M. Kutvirt, R.J. Waxweiller and C.R. Key, Uranium Mining and Lung Cancer in Navajo Men, *New England Journal of Medicine*, Vol.310, No.23, 7 June, 1984, pp.1481-1484.
I. Schmitz-Feuerhake and P. Carbonell, Evaluation of Low Level Effects in the Japanese A-Bomb Survivors After Current Dose Revisions and Estimation of Fallout Contribution, in *Biological Effects of Low-Level Radiation*, Proceedings of a Symposium, Venice, April, 1983, International Atomic Energy Agency, Vienna, 1983, pp.45-53.
I. Schmitz-Feuerhake, E. Muschol and K. Batjer, *Estimation of Somatic Risks and the Standards of the ICRP* , Paper presented to the Sixth International Congress of Radiation Research, Tokyo, 1979, 17 pages.
I. Schmitz-Feuerhake, E. Muschol, K. Batjer and R. Schafer, Risk Estimation of Radiation-Induced Thyroid Cancer in Adults, in *Late Biological Effects of Ionizing Radiation*, International Atomic Energy Agency, Vol.1, Vienna, 1978, pp.219-229.
G.G. Schofield, Leukaemia and Radiation, *The Lancet*, No.8146, 13 October, 1979, pp.802-803.

J. Sevc, E. Kunz and V. Placek, Lung Cancer in Uranium Miners and Long-Term Exposure to Radon Daughter Products, *Health Physics* , Vol.30, June, 1976, pp.433-437.

P.M.E. Sheehan and I.B. Hillary, An Unusual Cluster of Babies with Down's Syndrome Born to Former Pupils of an Irish Boarding School, *British Medical Journal* , Vol.287, No.6403, 12 November, 1983, pp.1428-1429.

R.E. Shore, R.E. Albert and B.S. Pasternack, Follow Up Study of Patients Treated by X-Ray Epilation for Tinea Capitis, *Archives of Environmental Health*, Vol.31, January/February 1976, pp.21-28.

Shutdown:Nuclear Power on Trial, The Book Publishing Co., Tennessee, 1979.

C. Silverman and D.A. Hoffman, Thyroid Tumor Risk from Radiation During Childhood, *Preventive Medicine*, Vol.4, 1975, pp.100-105.

J. Smith, *Clouds of Deceit: The Deadly Legacy of Britain's Bomb Tests* , Faber and Faber, 1985.

P.G. Smith, Some Problems in Assessing the Carcinogenic Risks to Man of Exposure to Ionizing Radiation, in *Energy and Health* edited by N.E. Breslow and A.S. Whittemore, Society for Industrial and Applied Mathematics, 1979, pp.61-80.

P.G. Smith and R. Doll, Mortality Among Patients with Ankylosing Spondylitis After a Single Treatment Course with X-Rays, *British Medical Journal*, Vol.284, 13 February 1982, pp.449-460.

R.J. Smith, Atom Bomb Tests Leave Infamous Legacy, *Science*, Vol.218, No.4569, 15 October, 1982, pp.266-269.

R.J. Smith, Study of Atomic Veterans Fuels Controversy, *Science*, Vol.221, No.4612, 19 August, 1983, pp.733-734.

V.K. Smith and C.C.S. Gilbert, The Implicit Valuation of Risks to Life: A Comparative Analysis, *Economics Letters*, Vol.16, No.3-4, 1984, pp.393-399.

A. Spackman and D. Connett, New Sellafield Scandal, *The Guardian*, 16 February, 1986.

F.W. Spiers and J. Vaughan, Hazards of Plutonium With Special Reference to the Skeleton, *Nature*, Vol.259, No.5544, 19 February, 1976, pp.531-534.

J. Stead, A-Bomb Test 'Cynicism' Documented, *The Guardian*, 28 November, 1984.

A.F. Stehney, H.F. Lucas and R.E. Rowland, Survival Times of Women Radium Dial Workers First Exposed Before 1930, in *Late Biological Effects of Ionizing Radiation*, International Atomic Energy Agency, Vol.1, Vienna, 1978, pp.333-351.

K. Stenstrand, M. Annanmaki and T. Rytomaa, Cytogenetric Investigation of People in Finland Using Household Water with High Natural Radioactivity, *Health Physics*, Vol.36, March, 1978, pp.441-444.

E.J. Sternglass, Environmental Radiation and Human Health, in *Proceedings of the Sixth Berkeley Symposium on Mathematical Statistics and Probability*, edited by L.M. Le Cam, J. Neyman and E. Scott, University of California Press, 1972, Vol.6, pp.145-221.

E.J. Sternglass, Nuclear Radiation and Human Health, in *Against Pollution and Hunger*, edited by A.M. Hilton, Universitetsforlaget, Oslo, 1974, pp.121-179.

E.J. Sternglass, Radioactivity, in *Environmental Chemistry*, edited by J.O.M. Bockris, Plenum Press, 1971, pp.477-515.

E.J. Sternglass, Health Effects of Low-Level Radiation, in *A Nuclear Ireland?*, edited by J.F. Carroll and P.K. Kelley, Irish TGWU, Dublin, 1978, pp.117-144.

E.J. Sternglass, *Secret Fallout: Low-Level Radiation from Hiroshima to Three Mile Island*, McGraw-Hill, 1981.

A.M. Stewart, Low Dose Radiation Cancers in Man, *Advances in Cancer Research*, Vol.14, 1971, pp.359-390.

A.M. Stewart, The Carcinogenic Effects of Low Level Radiation. A Re-Appraisal of Epidemiologists Methods and Observations, *Health Physics*, Vol.24, February, 1973a, pp.223-240.

A.M. Stewart, Cancer as a Cause of Abortions and Stillbirths; the Effect of these Early Deaths on the Recognition of Radiogenic Leukaemias, *British Journal of Cancer*, Vol.27, 1973b, pp.465-472.

A.M. Stewart, Low Dose Radiation: The Hanford Evidence, *The Lancet*, No.8072, 13 May, 1978a, pp.1048-1049.

A.M. Stewart, Low Dose Radiation, *The Lancet*, No.8076, 10 June, 1978b, p.1260.

A.M. Stewart, Proof of Evidence, in *Planning and Plutonium*, Town and Country Planning Association, 1978c, pp.68-81.

A.M. Stewart, Radiation Doses of Hanford Workers Dying from Various Causes, in *A Nuclear Ireland?*, edited by J.F. Carroll and P.K. Kelley, Irish TGWU, Dublin, 1978d, pp.145-158.

A.M. Stewart, Delayed Effects of A-Bomb Radiation: A Review of Recent Mortality Rates and Risk Estimates for Five-Year Survivors, *Journal of Epidemiology and Community Health* , Vol.26, No.2, June, 1982, pp.80-86.

A.M. Stewart and G.W. Kneale, Radiation Dose Effects in Relation to Obstetrical X-Rays and Childhood Cancers, *The Lancet*, No.7658, 6 June, 1970, pp.1185-1188.

A.M. Stewart, G.W. Kneale and T.F. Mancuso, The Hanford Data - A Reply to Recent Criticisms, *Ambio*, Vol.9, No.2, 1980, pp.66-73.

M. Stott and P.J. Taylor, *The Nuclear Controversy: A Guide to the Issues of the Windscale Inquiry*, Town and Country Planning Association and the Political Ecology Research Group, 1980.

Sunday Times Insight Team, *Rainbow Warrior: The French Attempt to Sink Greenpeace*, Arrow Books, 1986.

A.R. Tamplin and J.W. Gofman, The Radiation Effects Controversy, *The Bulletin of the Atomic Scientists*, Vol.26, No.2, September, 1970, p.2 and pp.5-8.

K.W. Taylor, N.L. Patt and H.E. Johns, Variations in X-Ray Exposures to Patients, *Journal of the Canadian Association of Radiologists*, Vol.30, No.1, March, 1979, pp.6-11.

P.J. Taylor, *The Windscale Fire, October, 1957*, Political Ecology Research Group, Oxford, Research Report RR-7, July, 1981.

P.J. Taylor, *The Impact of Nuclear Waste Disposal to the Marine Environment*, Political Ecology Research Group, Research Report RR-8, March 1982.

P.J. Taylor, *The Environmental Impact of a Projected Uranium Development in Co. Donegal, Ireland* , Political Ecology Research Group, Research Report, RR-9, 1983.

M.C. Thorne and J. Vennart, The Toxicity of ^{90}Sr, ^{226}Ra and ^{239}Pu, *Nature*, Vol.263, No.5578, 14 October, 1976, pp.555-558.

L. Torrey, Radiation Haunts Shipyard Workers, *New Scientist*, Vol.77, No.1094, 16 March, 1978, pp.726-727.
L. Torrey, Disease Legacy from Nevada Atomic Tests, *New Scientist*, Vol.84, No.1179, 1 November, 1979, pp.336-337.
L. Torrey, Nuclear Bomb Fall-Out Victims Should be Compensated, *New Scientist*, Vol.85, No.1199, 20 March, 1980a, p.901.
L. Torrey, Radiation Cloud Over Nuclear Power, *New Scientist*, Vol.86, No.1204, 24 April, 1980b, pp.197-199.
L. Torrey, Row About Nuclear Submarine Cancers, *New Scientist*, Vol.89, No.1234, 1 January, 1981, p.6.
A. Tucker, Inquiry Into Sizewell Deaths as Doctor Claims Leukaemia Link, *The Guardian*, 7 October, 1982.
A. Tucker, Detailed Reappraisal of Nuclear Test Victims Considered, *The Guardian*, 12 January, 1983a.
A. Tucker, Cloud of Unknowing, *The Guardian*, 17 March, 1983b.
A. Tucker, Seascale Warning Led to Sack for Scientist, *The Guardian*, 14 November, 1983c.
A. Tucker, Indecent Exposure, *The Guardian*, 24 November 1983d, p.23,
A. Tucker, Sellafield Leak is Referred to DPP, *The Guardian*, 22 December, 1983e.
A. Tucker, Figures Fuel N-Plant Leak Dispute, *The Guardian*, April, 1984.
A. Tucker, Why Shouldn't We Be Scared When They Won't Tell Us the Truth?, *The Guardian*, 1 March, 1986.
Y. Ujeno, Relation Between Cancer Incidence of Mortality and External Natural Background Radiation in Japan, in *Biological Effects of Low-Level Radiation*, Proceedings of a Symposium, Venice, April, 1983, International Atomic Energy Agency, Vienna, 1983, pp.253-262.
United Kingdom Atomic Energy Authority, *The Effects and Control of Radiation*, UKAEA, December, 1982.
UNSCEAR, *Sources and Effects of Ionizing Radiation*, United Nations Scientific Committee on the Effects of Atomic Radiation, 1977.
UNSCEAR, *Ionizing Radiation: Sources and Biological Effects* , United Nations Scientific Committee on the Effects of Atomic Radiation, 1982
A.C. Upton, Environmental Standards for Ionizing Radiation: Theoretical Basis for Dose-Response Curves, *Environmental Health Perspectives*, Vol.52, 1983, pp.31-39.
J. Urquhart, Polonium: Windscale's Most Lethal Legacy, *New Scientist*, Vol.97, No.1351, 31 March, 1983, pp.873-875.
J. Urquhart, J. Cutler and M. Burke, Leukaemia and Lymphatic Cancer in Young People Near Nuclear Installations, *The Lancet*, No.8477, 15 February, 1986, p.384.
I. Uzunov, F. Steinhausler and E. Pohl, Carcinogenic Risk of Exposure to Radon Daughters Associated with Radon Spas, *Health Physics*, Vol.41, No.6, December, 1981, pp.807-813.
G. Van Kaick, A. Kaul, D. Lorenz, M. Muth, K. Wegener and H. Wesch, Late Effects and Tissue Dose in Thorotrast Patients, in *Late Biological Effects of Ionizing Radiation*, International Atomic Energy Agency, Vienna, 1978, Vol.2, pp.263-276.
G.L. Voelz, What We Have Learned About Plutonium From Human Data, *Health Physics*, Vol.29, October, 1975, pp.551-561.

T. Wakabayashi, H. Kato, T.Ikeda and W.J. Schull, Studies of the Mortality of A-Bomb Survivors, Report 7: Part III. Incidence of Cancer in 1959-1978, Based on the Tumor Registry, Nagasaki, *Radiation Research*, Vol.93, 1983, pp.112-146.
R.W. Wallace and C.A. Sondhaus, Cosmic Radiation Exposure in Subsonic Air Transport, *Aviation, Space and Environmental Medicine*, Vol.49, No.4, April, 1978, pp.610-623.
A. Wardle, Sabotage of the Species, *The Observer*, 5 January, 1983, p.7.
J.P. Wesley, Background Radiation as the Cause of Fatal Congenital Malformation, *International Journal of Radiation Biology*, Vol.2, No.1, 1960, pp.97-118.
J. Wilkinson, Public Hazards and Curious News Value, *The Listener*, 31 May, 1984.
M. Wilkinson, Scientist Adrift on a Sea of Leaks, *The Financial Times* , 22 February, 1986.
S.M. Wolfe, Standards for Carcinogens: Science Affronted by Politics, in *Origins of Human Cancer*, edited by H.H. Hiatt, J.D. Watson and J.A. Winsten, Cold Spring Harbour Laboratory, Book C, 1977, pp.1735-1747.
R. Wood and N.D. Smith, Cancer Mortality Studied by Dounreay, *Atom*, No.357, July, 1986, pp.7-13.
A.D. Wrixon, Human Expense to Radon Decay Products, *Atom*, No.352, February, 1986, pp.2-6.
B. Wynne, The Politics of Nuclear Safety, *New Scientist*, Vol.77, No.1087, 26 January, 1978, pp.208-211.

Index